SpringerBriefs in Electrical and Computer Engineering

For further volumes:
http://www.springer.com/series/10059

Vishal M. Patel • Rama Chellappa

Sparse Representations and Compressive Sensing for Imaging and Vision

Vishal M. Patel
Center for Automation Research
University of Maryland
A.V. Williams Building
College Park, MD

Rama Chellappa
Department of Electrical and Computer
  Engineering and Center for
  Automation Research
A.V. Williams Building
University of Maryland
College Park, MD

ISSN 2191-8112   ISSN 2191-8120 (electronic)
ISBN 978-1-4614-6380-1   ISBN 978-1-4614-6381-8 (eBook)
DOI 10.1007/978-1-4614-6381-8
Springer New York Heidelberg Dordrecht London

Library of Congress Control Number: 2012956308

Printed on acid-free paper

Springer is part of Springer Science+Business Media (www.springer.com)

To my sisters Julie, Dharti and Gunjali
— Vishal M. Patel

Acknowledgements

We thank former and current students as well as collaborators - Richard Baraniuk, Volkan Cevher, Pavan Turaga, Ashok Veeraraghavan, Aswin Sankaranarayanan, Dikpal Reddy, Amit Agrawal, Nalini Ratha, Jaishanker Pillai, Hien Van Nguyen, Sumit Shekhar, Garrett Warnell, Qiang Qiu, Ashish Shrivastava - for letting us draw upon their work, thus making this monograph possible.

Research efforts summarized in this monograph were supported by the following grants and contracts: ARO MURI (W911NF-09-1-0383), ONR MURI (N00014-08-1-0638), ONR grant (N00014-12-1-0124), and a NIST grant (70NANB11H023).

Contents

Chapter 1
Introduction

Compressive sampling[1] [23, 47] is an emerging field that has attracted considerable interest in signal/image processing, computer vision and information theory. Recent advances in compressive sensing have led to the development of imaging devices that sense at measurement rates below than the Nyquist rate. Compressive sensing exploits the property that the sensed signal is often sparse in some transform domain in order to recover it from a small number of linear, random, multiplexed measurements. Robust signal recovery is possible from a number of measurements that is proportional to the sparsity level of the signal, as opposed to its ambient dimensionality.

While there has been remarkable progress in compressive sensing for static signals such as images, its application to sensing temporal sequences such as videos has also recently gained a lot of traction. Compressive sensing of videos makes a compelling application towards dramatically reducing sensing costs. This manifests itself in many ways including alleviating the data deluge problems [7] faced in the processing and storage of videos. Using novel sensors based on this theory, there is hope to accomplish tasks such as target tracking and object recognition while collecting significantly less data than traditional systems.

In this monograph, we will present an overview of the theories of sparse representation and compressive sampling and examine several interesting imaging modalities based on these theories. We will also explore the use of linear and non-linear kernel sparse representation as well as compressive sensing in many computer vision problems including target tracking, background subtraction and object recognition.

Writing this monograph presented a great challenge. Due to page limitations, we could not include all that we wished. We beg the forgiveness of many of our fellow researchers who have made significant contributions to the problems covered in this monograph and whose works could not be discussed.

---

[1] Also known as compressive sensing or compressed sensing.

V.M. Patel and R. Chellappa, *Sparse Representations and Compressive Sensing for Imaging and Vision*, SpringerBriefs in Electrical and Computer Engineering, DOI 10.1007/978-1-4614-6381-8_1, © The Author(s) 2013

1

1.1 Outline

We begin the monograph with a brief discussion on compressive sampling in Sect. 2. In particular, we present some fundamental premises underlying CS: sparsity, incoherent sampling and non-linear recovery. Some of the main results are also reviewed.

In Sect. 3, we describe several imaging modalities that make use of the theory of compressive sampling. In particular, we present applications in medical imaging, synthetic aperture radar imaging, millimeter wave imaging, single pixel camera and light transport sensing.

In Sect. 4, we present some applications of compressive sampling in computer vision and image understanding. We show how sparse representation and compressive sampling framework can be used to develop robust algorithms for target tracking. We then present several applications in video compressive sampling. Finally, we show how compressive sampling can be used to develop algorithms for recovering shapes and images from gradients.

Section 5 discusses some applications of sparse representation and compressive sampling in object recognition. In particular, we first present an overview of the sparse representation framework. We then show how it can be used to develop robust algorithms for object recognition. Through the use of Mercer kernels, we show how the sparse representation framework can be made non-linear. We also discuss multimodal multivariate sparse representation as well as its non-linear extension at the end of this section.

In Sect. 6, we discuss recent advances in dictionary learning. In particular, we present an overview of the method of optimal directions and the KSVD algorithms for learning dictionaries. We then show how dictionaries can be designed to achieve discrimination as well as reconstruction. Finally, we highlight some of the methods for learning non-linear kernel dictionaries.

Finally, concluding remarks are presented in Sect. 7.

Chapter 2
Compressive Sensing

Compressive sensing [47], [23] is a new concept in signal processing and information theory where one measures a small number of non-adaptive linear combinations of the signal. These measurements are usually much smaller than the number of samples that define the signal. From these small number of measurements, the signal is then reconstructed by a non-linear procedure. In what follows, we present some fundamental premises underlying CS: sparsity, incoherent sampling and non-linear recovery.

2.1 Sparsity

Let $\mathbf{x}$ be a discrete time signal which can be viewed as an $N \times 1$ column vector in $\mathbb{R}^N$. Given an orthonormal basis matrix $\mathbf{B} \in \mathbb{R}^{N \times N}$ whose columns are the basis elements $\{\mathbf{b}_i\}_{i=1}^N$, $\mathbf{x}$ can be represented in terms of this basis as

$$\mathbf{x} = \sum_{i=1}^N \alpha_i \mathbf{b}_i \tag{2.1}$$

or more compactly $\mathbf{x} = \mathbf{B}\alpha$, where α is an $N \times 1$ column vector of coefficients. These coefficients are given by $\alpha_i = \langle \mathbf{x}, \mathbf{b}_i \rangle = \mathbf{b}_i^T \mathbf{x}$ where $.^T$ denotes the transposition operation. If the basis $\mathbf{B}$ provides a K-sparse representation of $\mathbf{x}$, then (2.1) can be rewritten as

$$\mathbf{x} = \sum_{i=1}^K \alpha_{n_i} \mathbf{b}_{n_i},$$

where $\{n_i\}$ are the indices of the coefficients and the basis elements corresponding to the K nonzero entries. In this case, α is an $N \times 1$ column vector with only K nonzero elements. That is, $\|\alpha\|_0 = K$ where $\|.\|_p$ denotes the ℓ_p-norm defined as

V.M. Patel and R. Chellappa, *Sparse Representations and Compressive Sensing for Imaging and Vision*, SpringerBriefs in Electrical and Computer Engineering, DOI 10.1007/978-1-4614-6381-8_2, © The Author(s) 2013

$$\|\mathbf{x}\|_p = \left(\sum_i |x_i|^p \right)^{\frac{1}{p}}$$

and the ℓ_0-norm is defined as the limit as $p \to 0$ of the ℓ_p-norm

$$\|\mathbf{x}\|_0 = \lim_{p \to 0} \|\mathbf{x}\|_p^p = \lim_{p \to 0} \sum_i |x_i|^p.$$

In general, the ℓ_0-norm counts the number of non-zero elements in a vector

$$\|\mathbf{x}\|_0 = \sharp\{i : x_i \neq 0\}. \tag{2.2}$$

Typically, real-world signals are not exactly sparse in any orthogonal basis. Instead, they are *compressible*. A signal is said to be compressible if the magnitude of the coefficients, when sorted in a decreasing order, decays according to a power law [87],[19]. That is, when we rearrange the sequence in decreasing order of magnitude $\alpha_{(1)} \geq \alpha_{(2)} \geq \cdots \geq \alpha_{(N)}$, then the following holds

$$|\alpha|_{(n)} \leq C.n^{-s}, \tag{2.3}$$

where $|\alpha|_{(n)}$ is the nth largest entry of α, $s \geq 1$ and C is a constant. For a given L, the L-term linear combination of elements that best approximate $\mathbf{x}$ in an L_2-sense is obtained by keeping the L largest terms in the expansion

$$\mathbf{x}_L = \sum_{n=0}^{L-1} \alpha_{(n)} \mathbf{b}_{(n)}.$$

If α obeys (2.3), then the error between $\mathbf{x}_L$ and $\mathbf{x}$ also obeys a power law as well [87], [19]

$$\|\mathbf{x}_L - \mathbf{x}\|_2 \leq CL^{-(s-\frac{1}{2})}.$$

In other words, a small number of vectors from $\mathbf{B}$ can provide accurate approximations to $\mathbf{x}$. This type of approximation is often known as the *non-linear approximation* [87].

Fig. 2.1 shows an example of the non-linear approximation of the Boats image using Daubechies 4 wavelet. The original Boats image is shown in Fig. 2.1(a). Two level Daubechies 4 wavelet coefficients are shown in Fig. 2.1(b). As can be seen from this figure, these coefficients are very sparse. The plot of the sorted absolute values of the coefficients of the image is shown in Fig. 2.1(c). The reconstructed image after keeping only 10% of the coefficients with the largest magnitude is shown in Fig. 2.1(d). This reconstruction provides a very good approximation to the original image. In fact, it is well known that wavelets provide the best representation for piecewise smooth images. Hence, in practice wavelets are often used to compressively represent images.

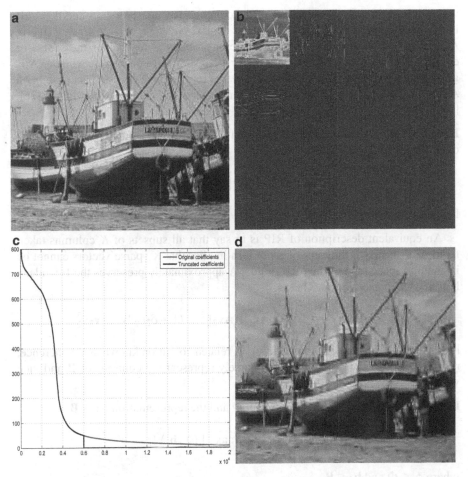

Fig. 2.1 Compressibility of wavelets. (**a**) Original Boats image. (**b**) Wavelet coefficients. (**c**) The plot of the sorted absolute values of the coefficients. (**d**) Reconstructed image after keeping only 10% of the coefficients with the largest magnitude

2.2 Incoherent Sampling

In CS, the K largest α_i in (2.1) are not measured directly. Instead, $M \ll N$ projections of the vector $\mathbf{x}$ with a collection of vectors $\{\phi_j\}_{j=1}^{M}$ are measured as in $y_j = \langle x, \phi_j \rangle$. Arranging the measurement vector ϕ_j^T as rows in an $M \times N$ matrix $\boldsymbol{\Phi}$ and using (2.1), the measurement process can be written as

$$\mathbf{y} = \boldsymbol{\Phi}\mathbf{x} = \boldsymbol{\Phi}\mathbf{B}\alpha = \mathbf{A}\alpha, \qquad (2.4)$$

where $\mathbf{y}$ is an $M \times 1$ column vector of the compressive measurements and $\mathbf{A} = \Phi\mathbf{B}$ is the measurement matrix or the sensing matrix. Given an $M \times N$ sensing matrix $\mathbf{A}$ and the observation vector $\mathbf{y}$, the general problem is to recover the sparse or compressible vector α. To this end, the first question is to determine whether $\mathbf{A}$ is good for compressive sensing. Candés and Tao introduced a necessary condition on $\mathbf{A}$ that guarantees a stable solution for both K sparse and compressible signals [26], [24].

Definition 2.1. A matrix $\mathbf{A}$ is said to satisfy the Restricted Isometry Property (RIP) of order K with constants $\delta_K \in (0, 1)$ if

$$(1 - \delta_K)\|\mathbf{v}\|_2^2 \leq \|\mathbf{A}\mathbf{v}\|_2^2 \leq (1 + \delta_K)\|\mathbf{v}\|_2^2$$

for any $\mathbf{v}$ such that $\|\mathbf{v}\|_0 \leq K$.

An equivalent description of RIP is to say that all subsets of K columns taken from $\mathbf{A}$ are nearly orthogonal. This in turn implies that K sparse vectors cannot be in the null space of $\mathbf{A}$. When RIP holds, $\mathbf{A}$ approximately preserves the Euclidean length of K sparse vectors. That is,

$$(1 - \delta_{2K})\|\mathbf{v_1} - \mathbf{v_2}\|_2^2 \leq \|\mathbf{A}\mathbf{v_1} - \mathbf{A}\mathbf{v_2}\|_2^2 \leq (1 + \delta_{2K})\|\mathbf{v_1} - \mathbf{v_2}\|_2^2$$

holds for all K sparse vectors $\mathbf{v_1}$ and $\mathbf{v_2}$. A related condition known as incoherence, requires that the rows of Φ can not sparsely represent the columns of $\mathbf{B}$ and vice versa.

Definition 2.2. The coherence between Φ and the representation basis $\mathbf{B}$ is

$$\mu(\Phi, \mathbf{B}) = \sqrt{N} \max_{1 \leq i,j \leq N} |\langle \phi_i, \mathbf{b}_j \rangle|, \qquad (2.5)$$

where $\phi_i \in \Phi$ and $\mathbf{b}_j \in \mathbf{B}$.

The number μ measures how much two vectors in $\mathbf{A} = \Phi\mathbf{B}$ can look alike. The value of μ is between 1 and $\sqrt{N}$. We say that a matrix $\mathbf{A}$ is *incoherent* when μ is very small. The incoherence holds for many pairs of bases. For example, it holds for the delta spikes and the Fourier bases. Surprisingly, with high probability, incoherence holds between any arbitrary basis and a random matrix such as Gaussian or Bernoulli [6], [142].

2.3 Recovery

Since, $M \ll N$, we have an under-determined system of linear equations, which in general has infinitely many solutions. So our problem is ill-posed. If one desires to narrow the choice to a well-defined solution, additional constraints are needed.

One approach is to find the minimum-norm solution by solving the following optimization problem

$$\hat{\alpha} = \arg\min_{\alpha'} \|\alpha'\|_2 \text{ subject to } \mathbf{y} = \mathbf{A}\alpha'.$$

The solution to the above problem is explicitly given by

$$\hat{\alpha} = \mathbf{A}^\dagger \mathbf{y} = \mathbf{A}^*(\mathbf{A}\mathbf{A}^*)^{-1}\mathbf{y},$$

where $\mathbf{A}^*$ is the adjoint of $\mathbf{A}$ and $\mathbf{A}^\dagger = \mathbf{A}^*(\mathbf{A}\mathbf{A}^*)^{-1}$ is the pseudo-inverse of $\mathbf{A}$. This solution, however, yields a non-sparse vector. The approach taken in CS is to instead find the sparsest solution.

The problem of finding the sparsest solution can be reformulated as finding a vector $\alpha \in \mathbb{R}^N$ with a minimum possible number of nonzero entries. That is

$$\hat{\alpha} = \arg\min_{\alpha'} \|\alpha'\|_0 \text{ subject to } \mathbf{y} = \mathbf{A}\alpha'. \tag{2.6}$$

This problem can recover a K sparse signal exactly. However, this is an NP-hard problem. It requires an exhaustive search of all $\binom{N}{K}$ possible locations of the nonzero entries in α.

The main approach taken in CS is to minimize the ℓ_1-norm instead

$$\hat{\alpha} = \arg\min_{\alpha'} \|\alpha'\|_1 \text{ subject to } \mathbf{y} = \mathbf{A}\alpha'. \tag{2.7}$$

Surprisingly, the ℓ_1 minimization yields the same result as the ℓ_0 minimization in many cases of practical interest. This program also approximates compressible signals. This convex optimization program is often known as Basis Pursuit (BP) [38]. The use of ℓ_1 minimization for signal restoration was initially observed by engineers working in seismic exploration as early as 1970s [52]. In the last few years, a series of papers [47], [142], [21], [25], [19], [22], explained why ℓ_1 minimization can recover sparse signals in various practical setups.

2.3.1 Robust CS

In this section we examine the case when there are noisy observations of the following form

$$\mathbf{y} = \mathbf{A}\alpha + \eta \tag{2.8}$$

where $\eta \in \mathbb{R}^M$ is the measurement noise or an error term. Note that η can be stochastic or deterministic. Furthermore, let's assume that $\|\eta\|_2 \leq \varepsilon$. Then, $\mathbf{x}$ can be recovered from $\mathbf{y}$ via α by solving the following problem

$$\hat{\alpha} = \arg\min_{\alpha'} \|\alpha'\|_1 \text{ subject to } \|\mathbf{y} - \mathbf{A}\alpha'\| \leq \varepsilon. \tag{2.9}$$

The problem (2.9) is often known as Basis Pursuit DeNoising (BPDN) [38]. In [22], Candés *at. el.* showed that the solution to (2.9) recovers an unknown sparse signal with an error at most proportional to the noise level.

Theorem 2.1. *[22] Let* $\mathbf{A}$ *satisfy RIP of order 4K with* $\delta_{3K} + 3\delta_{4K} < 2$. *Then, for any K sparse signal* α *and any perturbation* η *with* $\|\eta\|_2 \leq \varepsilon$, *the solution* $\hat{\alpha}$ *to (2.9) obeys*

$$\|\hat{\alpha} - \alpha\|_2 \leq \varepsilon C_K$$

with a well behaved constant C_K.

Note that for K obeying the condition of the theorem, the reconstruction from noiseless data is exact. A similar result also holds for stable recovery from imperfect measurements for approximately sparse signals (i.e compressible signals).

Theorem 2.2. *[22] Let* $\mathbf{A}$ *satisfy RIP of order 4K. Suppose that* α *is an arbitrary vector in* $\mathbb{R}^N$ *and let* α_K *be the truncated vector corresponding to the K largest values of* θ *in magnitude. Under the hypothesis of Theorem 2.1, the solution* $\hat{\alpha}$ *to (2.9) obeys*

$$\|\hat{\alpha} - \alpha\|_2 \leq \varepsilon C_{1,K} + C_{2,K} \frac{\|\alpha - \alpha_K\|_1}{\sqrt{K}}$$

with well behaved constants $C_{1,K}$ *and* $C_{2,K}$.

If α obeys (2.3), then

$$\frac{\|\hat{\alpha} - \alpha_K\|_1}{\sqrt{K}} \leq C'K^{-(s-\frac{1}{2})}.$$

So in this case

$$\|\hat{\alpha} - \alpha_K\|_2 \leq C''K^{-(s-\frac{1}{2})},$$

and for signal obeying (2.3), there are fundamentally no better estimates available. This, in turn, means that with only M measurements, one can achieve an approximation error which is almost as good as that one obtains by knowing everything about the signal α and selecting its K-largest elements [22].

2.3.1.1 The Dantzig selector

In (2.8), if the noise is assumed to be Gaussian with mean zero and variance σ^2, $\eta \sim \mathcal{N}(0, \sigma^2)$, then the stable recovery of the signal is also possible by solving a modified optimization problem

$$\hat{\alpha} = \arg\min_{\alpha'} \|\alpha'\|_1 \text{ s. t. } \|\mathbf{A}^T(\mathbf{y} - \mathbf{A}\alpha')\|_\infty \leq \varepsilon' \tag{2.10}$$

where $\varepsilon' = \lambda_N \sigma$ for some $\lambda_N > 0$ and $\|.\|_\infty$ denotes the ℓ_∞ norm. For an N dimensional vector $\mathbf{x}$, it is defined as $\|\mathbf{x}\|_\infty = \max(|x_1|, \cdots, |x_N|)$. The above program is known as the Dantzig Selector [28].

Theorem 2.3. *[28] Suppose $\alpha \in \mathbb{R}^N$ is any K-sparse vector obeying $\delta_{2K} + \vartheta_{K,2K} < 1$. Choose $\lambda_N = \sqrt{2\log(N)}$ in (2.10). Then, with large probability, the solution to (2.10), $\hat{\alpha}$ obeys*

$$\|\hat{\alpha} - \alpha\|_2^2 \leq C_1^2 \cdot (2\log(N)) \cdot K \cdot \sigma^2, \tag{2.11}$$

with

$$C_1 = \frac{4}{1 - \delta_K - \vartheta_{K,2K}},$$

where $\vartheta_{K,2K}$ is the $K, 2K$-restricted orthogonal constant defined as follows

Definition 2.3. The K, K'-restricted orthogonality constant $\vartheta_{K,K'}$ for $K + K' \leq N$ is defined to be the smallest quantity such that

$$|\langle \mathbf{A}_T \mathbf{v}, \mathbf{A}_{T'} \mathbf{v}' \rangle| \leq \vartheta_{K,K'} \|\mathbf{v}\|_2 \|\mathbf{v}'\|_2 \tag{2.12}$$

holds for all disjoint sets $T, T' \subseteq \{1, ..., N\}$ of cardinality $|T| \leq K$ and $|T'| \leq K'$.

A similar result also exists for compressible signals (see [28] for more details).

2.3.2 CS Recovery Algorithms

The ℓ_1 minimization problem (2.10) is a linear program [28] while (2.9) is a second-order cone program (SOCP) [38]. SOCPs can be solved using interior point methods [74]. Log-barrier and primal dual methods can also be used [15], [3] to solve SOCPs. Note, the optimization problems (2.7), (2.9), and (2.10) minimize convex functionals, hence a global minimum is guaranteed.

In what follows, we describe other CS related reconstruction algorithms.

2.3.2.1 Iterative Thresholding Algorithms

A Lagrangian formulation of the problem (2.9) is the following

$$\hat{\alpha} = \arg\min_{\alpha'} \|\mathbf{y} - \mathbf{A}\alpha'\|_2^2 + \lambda \|\alpha'\|_1. \tag{2.13}$$

There exists a mapping between λ from (2.13) and ε from (2.9) so that both problems (2.9) and (2.13) are equivalent. Several authors have proposed to solve (2.13) iteratively [12], [45], [11], [9]. This algorithm iteratively performs a soft-thresholding to decrease the ℓ_1 norm of the coefficients α and a gradient descent to decrease the value of $\|\mathbf{y} - \mathbf{A}\alpha\|_2^2$. The following iteration is usually used

$$\mathbf{y}^{n+1} = T_\lambda \left(\mathbf{y}^n + \mathbf{A}^*(\alpha - \mathbf{A}\mathbf{y}^n) \right), \tag{2.14}$$

where T_λ is the element wise soft-thresholding operator

$$T_\lambda(a) = \begin{cases} a + \frac{\lambda}{2}, & \text{if } a \le \frac{-\lambda}{2} \\ 0, & \text{if } |a| < \frac{\lambda}{2} \\ a - \frac{\lambda}{2}, & \text{if } a \ge \frac{\lambda}{2}. \end{cases}$$

The iterates y^{n+1} converge to the solution of (2.9), $\hat{\alpha}$ if $\|A\|_2 < 1$ [45]. Similar results can also be obtained using the hard-thresholding instead of the soft-thresholding in (2.14) [11].

Other methods for solving (2.13) have also been proposed. See for instance GPSR [61], SPGL1 [8], Bregman iterations [159], split Bregman iterations [65], SpaRSA [157], and references therein.

2.3.2.2   Greedy Pursuits

In certain conditions, greedy algorithms such as matching pursuit [88], orthogonal matching pursuit [109], [138], gradient pursuits [13], regularized orthogonal matching pursuit [94] and Stagewise Orthogonal Matching Pursuit [49] can also be used to recover sparse (or in some cases compressible) α from (2.8). In particular, a greedy algorithm known as, CoSaMP, is well supported by theoretical analysis and provides the same guarantees as some of the optimization based approaches [93].

Let T be a subset of $\{1, 2, \cdots, N\}$ and define the restriction of the signal x to the set T as

$$x|_T = \begin{cases} x_i, & i \in T \\ 0, & \text{otherwise} \end{cases}$$

Let A_T be the column submatrix of A whose columns are listed in the set T and define the pseudoinverse of a tall, full-rank matrix C by the formula $C^\ddagger = (C^*C)^{-1}C^*$. Let $supp(x) = \{x_j : j \ne 0\}$. Using this notation, the pseudo-code for CoSaMP is given in Algorithm 1 which can be used to solve the under-determined system of linear equations (2.4).

2.3.2.3   Other Algorithms

Recently, there has been a great interest in using ℓ_p minimization with $p < 1$ for compressive sensing [37]. It has been observed that the minimization of such a nonconvex problem leads to recovery of signals that are much less sparse than required by the traditional methods [37].

Other related algorithms such as FOCUSS and reweighted ℓ_1 have also been proposed in [68] and [29], respectively.

Algorithm 1: Compressive Sampling Matching Pursuit (CoSaMP)
Input: A, y, sparsity level K.
Initialize: $\alpha_0 = \mathbf{0}$ and the current residual $\mathbf{r} = \mathbf{y}$.
While not converged do
1. Compute the current error:
$\mathbf{v} = \mathbf{A}^*\mathbf{r}$.
2. Compute the best 2K support set of the error:
$\Omega = \mathbf{v}_{2K}$.
3. Merge the the strongest support sets:
$T = \Omega \cup supp(\alpha_{J-1})$.
4. Perform a least-squares signal estimation:
$\mathbf{b}_{|_T} = \mathbf{A}_{|_T}^{\ddagger}\mathbf{y}, \mathbf{b}_{T^c} = \mathbf{0}$.
5. Prune α_{J-1} and compute $\mathbf{r}$ for the next round:
$\alpha_J = \mathbf{b}_k, \mathbf{r} = \mathbf{y} - \mathbf{A}\alpha_J$.

2.4 Sensing Matrices

Most of the sensing matrices in CS are produced by taking i.i.d. random variables with some given probability distribution and then normalizing their columns. These matrices are guaranteed to perform well with high probability. In what follows, we present some commonly used sensing matrices in CS [22], [142], [26].

- *Random matrices with i.i.d. entries:* Consider a matrix **A** with entries drawn independently from the Gaussian probability distribution with mean zero and variance $1/M$. Then the conditions for Theorem 2.1 hold with overwhelming probability when

$$K \leq CM / \log(N/M).$$

- *Fourier ensemble:* Let **A** be an $M \times N$ matrix obtained by selecting M rows, at random, from the $N \times N$ discrete Fourier transform matrix and renormalizing the columns. Then with overwhelming probability, the conditions for Theorem 2.1 holds provided that

$$K \leq C \frac{M}{(\log(N))^6}.$$

- *General orthogonal ensembles:* Suppose **A** is obtained by selecting M rows from an $N \times N$ orthonormal matrix Θ and renormalizing the columns. If the rows are selected at random, then the conditions for Theorem 2.1 hold with overwhelming probability when

$$K \leq C \frac{1}{\mu^2} \frac{M}{(\log(N))^6},$$

where μ is defined in (2.5).

2.5 Phase Transition Diagrams

The performance of a CS system can be evaluated by generating phase transition
diagrams [86], [48], [10], [51]. Given a particular CS system, governed by the
sensing matrix $\mathbf{A} = \boldsymbol{\Phi}\mathbf{B}$, let $\delta = \frac{M}{N}$ be a normalized measure of undersampling
factor and $\rho = \frac{K}{M}$ be a normalized measure of sparsity. A plot of the pairing of the
variables δ and ρ describes a 2-D phase space $(\delta, \rho) \in [0, 1]^2$. It has been shown that
for many practical CS matrices, there exist sharp boundaries in this phase space that
clearly divide the solvable from unsolvable problems in the noiseless case. In other
words, a phase transition diagram provides a way of checking ℓ_0/ℓ_1 equivalence,
indicating how sparsity and indeterminacy affect the success of ℓ_1 minimization
[86], [48], [51]. Fig. 2.2 shows an example of a phase transition diagram which is
obtained when a random Gaussian matrix is used as $\mathbf{A}$. Below the boundary, ℓ_0/ℓ_1
equivalence holds and above the boundary, the system lacks sparsity and/or too few
measurements are obtained to solve the problem correctly.

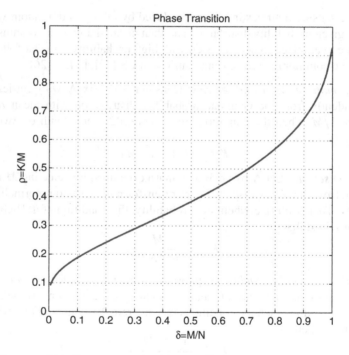

Fig. 2.2 Phase transition diagram corresponding to a CS system where $\mathbf{A}$ is the random Gaussian
matrix. The boundary separates regions in the problem space where (2.7) can and cannot be solved.
Below the curve solutions can be obtained and above the curve solutions can not be obtained

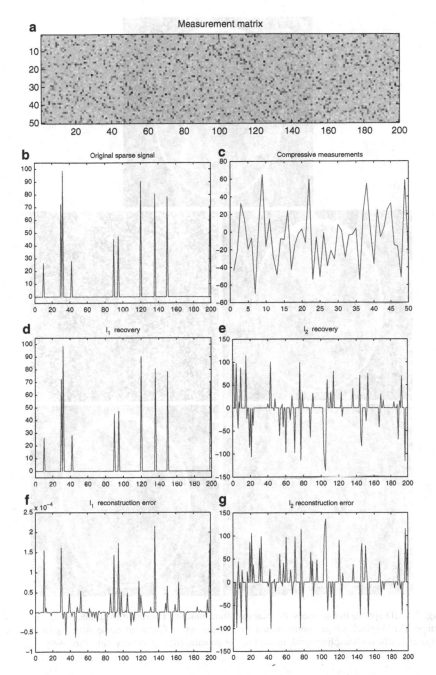

Fig. 2.3 1D sparse signal recovery example from random Gaussian measurements. (**a**) Compressive measurement matrix. (**b**) Original sparse signal. (**c**) Compressive measurements. (**d**) ℓ_1 recovery. (**e**) ℓ_2 recovery. (**f**) ℓ_1 reconstruction error. (**g**) ℓ_2 reconstruction error

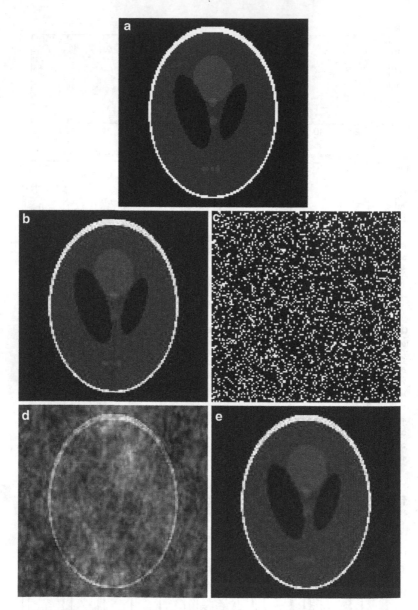

Fig. 2.4 2D sparse image recovery example from random Fourier measurements. (**a**) Original image. (**b**) Original image contaminated by additive white Gaussian noise with signal-to-noise ratio of 20 dB. (**c**) Sampling mask in the Fourier domain. (**d**) ℓ_2 recovery. (**e**) ℓ_1 recovery

2.6 Numerical Examples

We end this section by considering the following two examples. In the first example, a 1D signal $\mathbf{x}$ of length 200 with only 10 nonzero elements is undersampled using a random Gaussian matrix Φ of size 50×200 as shown in Fig. 2.3(a). Here, the sparsifying transform $\mathbf{B}$ is simply the identity matrix and the observation vector $\mathbf{y}$ is of length 50. Having observed $\mathbf{y}$ and knowing $\mathbf{A} = \Phi$ the signal $\mathbf{x}$ is then recovered by solving the following optimization problem

$$\hat{\mathbf{x}} = \arg \min_{\mathbf{x}' \in \mathbb{R}^N} \|\mathbf{x}'\|_1 \text{ subject to } \mathbf{y} = \mathbf{A}\mathbf{x}'. \tag{2.15}$$

As can be seen from Fig. 2.3(d), indeed the solution to the above optimization problem recovers the sparse signal exactly from highly undersampled observations. Whereas, the minimum norm solution (i.e. by minimizing the ℓ_2 norm), as shown in Fig. 2.3(e), fails to recover the sparse signal. The errors corresponding the ℓ_1 and ℓ_2 recovery are shown in Fig. 2.3(f) and Fig. 2.3(g), respectively.

In the second example, we reconstructed an undersampled Shepp-Logan phantom image of size 128×128 in the presence of additive white Gaussian noise with signal-to-noise ratio of 30 dB. For this example, we used only 15% of the random Fourier measurements and Haar wavelets as a sparsifying transform. So the observations can be written as $\mathbf{y} = \mathbf{MFB}\alpha + \eta$, where $\mathbf{y}, \mathbf{M}, \mathbf{F}, \mathbf{B}, \alpha$ and η are the noisy compressive measurements, the restriction operator, Fourier transform operator, the Haar transform operator, the sparse coefficient vector and the noise vector with $\|\eta\|_2 \leq \varepsilon$, respectively. The image was reconstructed via α estimated by solving the following optimization problem

$$\hat{\alpha} = \arg \min_{\alpha'} \|\alpha'\|_1 \text{ subject to } \|\mathbf{y} - \mathbf{MFB}\alpha'\| \leq \varepsilon.$$

The reconstruction from ℓ_2 and ℓ_1 minimization is shown in Fig. 2.4(d) and Fig. 2.3(e), respectively. This example shows that, it is possible to obtain a stable reconstruction from the compressive measurements in the presence of noise. For both of the above examples we used SPGL1 [8] algorithm for solving the ℓ_1 minimization problems.

In [23], [47], a theoretical bound on the number of samples that need to be measured for a good reconstruction has been derived. However, it has been observed by many researchers [79], [22], [142], [19], [26] that in practice samples in the order of two to five times the number of sparse coefficients suffice for a good reconstruction. Our experiments also support this claim.

Chapter 3
Compressive Acquisition

Many imaging modalities have been proposed that make use of the theory of compressive sensing. In this chapter, we present several sensors designed using CS theory. In particular, we focus on the Single Pixel Camera (SPC) [54], [149], Magnetic Resonance Imaging (MRI) [79], [80], [108], Synthetic Aperture Radar (SAR) imaging [103], passive millimeter wave imaging [104] and compressive light transport sensing [110]. See [153] and [55] for excellent tutorials on the applications of compressive sensing in the context of optical imaging as well as analog-to-information conversion.

3.1 Single Pixel Camera

One of the first physical imagers that demonstrated the practicality of compressive sensing in imaging was the Rice single pixel camera [54], [149]. The SPC essentially measures the inner products between an N-pixel sampled version of the incident light-field from the scene and a set of N-pixel test functions to capture the compressive measurements. The SPC architecture is illustrated in Fig. 3.1.

This architecture uses only a single detector element to image the scene. A digital micromirror array is used to represent a pesudorandom binary array. The light-field is then projected onto that array and the intensity of the projection is measured with a single detector. The orientations of the mirrors in the micromirror array can be changed rapidly, as a result a series of different pseudorandom projections can be measured in relatively little time. The scene is then reconstructed from compressive measurements using the existing CS reconstruction algorithms [54], [149].

Sample image reconstructions from SPC are shown in Fig. 3.2. A black-and-white picture of an "R" is used to reconstruct using SPC. The original dimension of the image is $N = 256 \times 256$. The reconstructed images using total variation minimization from only 2% and 10% measurements are shown in the second and third columns of Fig. 3.2, respectively.

V.M. Patel and R. Chellappa, *Sparse Representations and Compressive Sensing for Imaging and Vision*, SpringerBriefs in Electrical and Computer Engineering, DOI 10.1007/978-1-4614-6381-8_3, © The Author(s) 2013

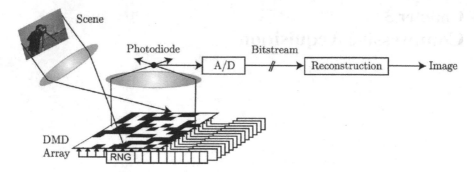

Fig. 3.1 Single pixel camera block diagram [149]

Fig. 3.2 Sample image reconstructions. (a) 256 × 256 original image. (b) Image reconstructed from only 2% of the measurements. (c) Image reconstructed from only 10% of the measurements [149]

One of the main limitations of this architecture is that it requires the camera to be focused on the object of interest until enough measurements are collected. This may be prohibitive in some applications.

3.2 Compressive Magnetic Resonance Imaging

Magnetic resonance imaging is based on the principle that protons in water molecules in the human body align themselves in a magnetic field. In an MRI scanner, radio frequency fields are used to systematically alter the alignment of the magnetization. This causes the nuclei to produce a rotating magnetic field which is recorded to construct an image of the scanned area of the body. Magnetic field gradients cause nuclei at different locations to rotate at different speeds. By using gradients in different directions 2D images or 3D voxels can be imaged in any arbitrary orientation. The magnetic field measured in an MRI scanner corresponds to Fourier coefficients of the imaged objects. The image is then recovered by taking the inverse Fourier transform. In this way, one can view an MRI scanner as a machine that measures the information about the object in Fourier domain. See [154] for an excellent survey on MRI.

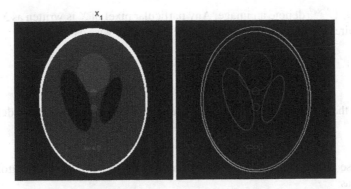

Fig. 3.3 512 × 512 Shepp-Logan Phantom image and its edges

One of the major limitations of the MRI is the linear relation between the number of measured data and scan time. As a result MRI machines tend to be slow, claustrophobic, and generally uncomfortable for patients. It would be beneficial for patients if one could significantly reduce the number of measurements that these devices take in order to generate a high quality image. Hence, methods capable of reconstructing from such partial sample sets would greatly reduce a patient's exposure time.

The theory of CS can be used to reduce the scan time in MRI acquisition by exploiting the transform domain sparsity of the MR images [79], [80], [108]. The standard techniques result in aliasing artifacts when the partial Fourier measurements are acquired. However, using sparsity as a prior in the MR images, one can reconstruct the image using the sparse recovery methods without the aliasing artifacts. While most MR images are not inherently sparse, they are sparse with respect to the total variation (TV). Most CS approaches to recovering such images utilize convex programs similar to that of Basis Pursuit. Instead of minimizing the ℓ_1 norm of the image subject to Fourier constraints, such programs minimize the TV semi-norm which enforces the necessary total variational sparsity of the solution (see [23,35,151], and others). While this methodology yields a significant reduction in the number of Fourier samples required to recover a sparse-gradient image, it does not take advantage of additional sparsity that can be exploited by utilizing the two horizontal and vertical directional derivatives of the image. An example of a sparse-gradient image along with an image of its edges is shown in Fig. 3.3.

An interesting approach to the problem of recovering a sparse gradient image from a small set of Fourier measurements was proposed in [108]. By using the fact the Fourier transform of the gradients of an image are precisely equal to a diagonal transformation of the Fourier transform of the original image, they utilize CS methods to directly recover the horizontal and vertical differences of the desired image. Then, integration techniques are performed to recover the original image from the edge estimates.

Let $X \in \mathbb{C}^{N \times N}$ denote an image. Any particular pixel of X is written as $X_{n,m}$. The discrete directional derivatives on X are defined pixel-wise as

$$(X_x)_{n,m} = X_{n,m} - X_{n-1,m}$$

$$(X_y)_{n,m} = X_{n,m} - X_{n,m-1}.$$

Based on these, the discrete gradient operator ∇ where $\nabla X \in \mathbb{C}^{N \times N \times 2}$ is defined as

$$(\nabla X)_{n,m} = ((X_x)_{n,m}, (X_y)_{n,m}).$$

From these operators, one can define the discrete total-variational operator TV or $|\nabla|$ on X as

$$(TV[X])_{n,m} = (|\nabla|(X))_{n,m}$$

$$= \sqrt{|(X_x)_{n,m}|^2 + |(X_y)_{n,m}|^2}, \tag{3.1}$$

from which one can also define the total-variation seminorm of X as

$$\|X\|_{TV} = \|TV(X)\|_1.$$

It is said that X is K-sparse in gradient (or in the total-variational sense) if $\||\nabla|(X)\|_0 = K$.

The objective is to recover an image X that is K-sparse in gradients from a set of $M \ll N^2$ Fourier measurements. To that end, define a set Ω of M two-dimensional frequencies $\underline{\omega}_k = (\omega_{x,k}, \omega_{y,k})$, $1 \leq k \leq M$ chosen according to a particular sampling pattern from $\{0, 1, \cdots, N-1\}^2$. Let $\mathscr{F}$ denote the two-dimensional DFT of X

$$\mathscr{F}(\omega_x, \omega_y) = \sum_{n=0}^{N-1} \sum_{m=0}^{N-1} X(n,m) \exp\left(-2\pi i \left(\frac{n\omega_x}{N}, \frac{m\omega_y}{N}\right)\right)$$

and $\mathscr{F}^{-1}$ its inverse

$$\mathscr{F}^{-1}\{\mathscr{F}(\omega_x, \omega_y)\} = X(n,m) = \frac{1}{N^2} \sum_{\omega_x=0}^{N-1} \sum_{\omega_y=0}^{N-1} \mathscr{F}(\omega_x, \omega_y) \exp\left(2\pi i \left(\frac{n\omega_x}{N}, \frac{m\omega_y}{N}\right)\right).$$

Next define the operator $\mathscr{F}_\Omega : \mathbb{C}^{N \times N} \to \mathbb{C}^M$ as

$$(\mathscr{F}_\Omega X)_k = (\mathscr{F}X)_{\underline{\omega}_k} \tag{3.2}$$

i.e. Fourier transform operator restricted to Ω. $\mathscr{F}_\Omega^*$ will represent its conjugate adjoint. Equipped with the above notation, the main problem considered in [108] can be formally stated as follows:

3.1. Given a set Ω of $M \ll N^2$ frequencies and Fourier observations of a K-sparse in gradient image X given by $\mathscr{F}_\Omega X$, how can one estimate X accurately and efficiently?

The most popular method of solving this problem is to find the image of least total variation that satisfies the given Fourier constraints. This corresponds to solving the following convex optimization problem

$$\tilde{X} = \underset{Y}{\arg\min} \|Y\|_{TV} \text{ s.t. } \mathscr{F}_{\Omega} Y = \mathscr{F}_{\Omega} X \qquad (3.3)$$

Based on an extension of Theorem 1.5 in [23] and the result in [122] regarding Fourier measurements, one can prove the following proposition.

Proposition 3.1 *Let X be a real-valued K-sparse in gradient image. If $M = \mathscr{O}(K \log^4 N)$, then the solution $\tilde{X}$ of (3.3) is unique and equal to X with probability at least $1 - \mathscr{O}(N^{-M})$.*

In the case of an image corrupted by noise, the measurements take the form

$$b = \mathscr{F}_{\Omega} X + \eta \qquad (3.4)$$

where η is the measurement noise with $\|\eta\|_2 = \varepsilon$. This problem can be solved by a similar convex optimization problem, which can be written as:

$$\tilde{X} = \underset{Y}{\arg\min} \|Y\|_{TV} \text{ s.t. } \|\mathscr{F}_{\Omega} Y - b\|_2 \leq \varepsilon. \qquad (3.5)$$

Several other methods have also been proposed that make use of the TV norm for recovering images from compressive measurements [35], [151], [143], [79], [158], [91], [70], [141], [159].

It was shown in [108] that instead of reconstructing an image by TV minimization, one can reconstruct the image by separately reconstructing the gradients and then solving for the image. This allows one to reconstruct the image with a far fewer number of measurements than required by the TV minimization method. Figure 3.4 presents an important comparison in the sparsity of X_x, X_y and the TV measure. The plots of the sorted absolute values of the coefficients of the gradients X_x, X_y and the TV measure for the Shepp-Logan Phantom image (Fig. 3.3) indicate that X_x and X_y decay much faster than the TV measure. In fact, it is easy to see from the expression of TV (equation (3.1)) that the coefficients of X_x and X_y will always decay faster than the coefficients of TV. This means, one can take advantage of this and be able to reconstruct an image with far fewer measurements than that required by using the TV-based method.

3.2.1 Image Gradient Estimation

Given Fourier observations $\mathscr{F}_{\Omega} X$ over some set of frequencies Ω, one can obtain the Fourier observations of X_x and X_y over Ω via the equations:

$$(\mathscr{F}_{\Omega} X_x)_k = (1 - e^{-2\pi i \omega_{x,k}/N})(\mathscr{F}_{\Omega} X)_k \qquad (3.6)$$

$$(\mathscr{F}_{\Omega} X_y)_k = (1 - e^{-2\pi i \omega_{y,k}/N})(\mathscr{F}_{\Omega} X)_k. \qquad (3.7)$$

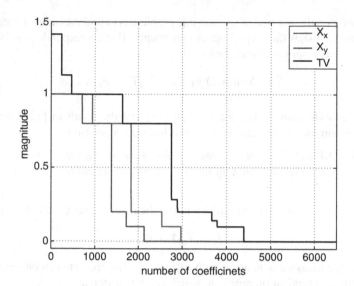

Fig. 3.4 The magnitude of *TV* (black), X_x (red) and X_y (blue) coefficients in decreasing order for the Shepp-Logan Phantom image (see Figure 3.3)

After this is done, any one of many CS recovery algorithms can be used to recover X_x and X_y from their respective Fourier observations. For instance, taking into account the presence of additive noise during the measurement process, gradients can be estimated by solving the following two optimization problems:

$$\widetilde{X}_x = \arg\min_{X_x'} \|X_x'\|_1 \text{ s. t. } \|\mathscr{F}_\Omega X_x' - b_x\|_2 \le \varepsilon_x \tag{3.8}$$

$$\widetilde{X}_y = \arg\min_{X_y'} \|X_y'\|_1 \text{ s. t. } \|\mathscr{F}_\Omega X_y' - b_y\|_2 \le \varepsilon_y \tag{3.9}$$

where we have assumed that the measurements are of the following form

$$b_x = \mathscr{F}_\Omega X_x + \eta_x$$

$$b_y = \mathscr{F}_\Omega X_y + \eta_y$$

with $(\eta_x)_k = (1 - e^{-2\pi i \omega_{x,k}/N})(\eta)_k$, $\varepsilon_x = \|\eta_x\|_2$, $(\eta_y)_k = (1 - e^{-2\pi i \omega_{y,k}/N})(\eta)_k$, and $\varepsilon_y = \|\eta_y\|_2$.

Note that the recovery of sparse gradients from their respective Fourier measurements will depend on the RIP of the resulting sensing matrix. It is very difficult to prove that the resulting CS matrix satisfies the RIP for any particular restriction set Ω. However, empirical studies have shown that RIP holds for many practical measurement schemes arising in medical imaging [23], [79], [141].

3.2.2 Image Reconstruction from Gradients

After obtaining estimates $\widetilde{X}_x$ and $\widetilde{X}_y$ of X_x and X_y, respectively, some kind of integration must be performed to recover an estimate $\check{X}$ of X. To obtain X from $\widetilde{X}_x$ and $\widetilde{X}_y$, the following optimization problem was proposed in [108]:

$$\check{X} = \arg\min_{Y} \left\| Y_x - \widetilde{X}_x \right\|_2^2 + \left\| Y_y - \widetilde{X}_y \right\|_2^2 + \beta \left\| Y_x \right\|_2^2 + \beta \left\| Y_y \right\|_2^2 + \lambda \left\| \mathscr{F}_\Omega Y - b \right\|_2^2.$$

where β and λ are penalty parameters that determine the degrees to which the TV-minimization and Fourier constraints are enforced.

Now observe that if hats are used to denote the Fourier Transform operator, it is possible to use the Parseval's Theorem to rewrite (3.10) as the following equivalent problem in the Fourier domain:

$$\hat{\check{X}} = \arg\min_{\hat{Y}} \left\| \left(1 - e^{-2\pi i \omega_1/N}\right)\hat{Y} - \hat{\widetilde{X}}_x \right\|_2^2 + \left\| \left(1 - e^{-2\pi i \omega_2/N}\right)\hat{Y} - \hat{\widetilde{X}}_y \right\|_2^2$$

$$+ \beta\left(\left\| \left(1 - e^{-2\pi i \omega_1/N}\right)\hat{Y} \right\|_2^2\right) + \beta\left(\left\| \left(1 - e^{-2\pi i \omega_2/N}\right)\hat{Y} \right\|_2^2\right)$$

$$+ \lambda \left\| (\hat{Y} - B)\mathbf{1}_\Omega \right\|_2^2. \tag{3.10}$$

Here $\mathbf{1}_\Omega$ denotes an indicator function which is 1 on Ω and 0 otherwise. Similarly, B is an $N \times N$ matrix that is equal to b on Ω and 0 otherwise. Based on this convenient alternative formulation of the problem, the following result was derived in [108]:

Proposition 3.1. *The least squares problem (3.10) can be solved element-wise by (3.11). Furthermore, if one lets $\lambda \to \infty$, then this solution will take the piecewise form (3.12).*

$$\hat{\check{X}}_{\omega_1,\omega_2} = \frac{\left(1 - e^{2\pi i \omega_1/N}\right)(\hat{\widetilde{X}}_x)_{\omega_1,\omega_2} + \left(1 - e^{2\pi i \omega_2/N}\right)(\hat{\widetilde{X}}_y)_{\omega_1,\omega_2} + \lambda B_{\omega_1,\omega_2}\mathbf{1}_\Omega}{(1+\beta)\left(\left|1 - e^{-2\pi i \omega_1/N}\right|^2 + \left|1 - e^{-2\pi i \omega_2/N}\right|^2\right) + \lambda\mathbf{1}_\Omega}. \tag{3.11}$$

$$\hat{\check{X}}_{\omega_1,\omega_2} = \begin{cases} B_{\omega_1,\omega_2} & \text{if } (\omega_1,\omega_2) \in \Omega \\ \frac{\left(1-e^{2\pi i \omega_1/N}\right)(\hat{\widetilde{X}}_x)_{\omega_1,\omega_2} + \left(1-e^{2\pi i \omega_2/N}\right)(\hat{\widetilde{X}}_y)_{\omega_1,\omega_2}}{(1+\beta)\left(\left|1-e^{-2\pi i \omega_1/N}\right|^2 + \left|1-e^{-2\pi i \omega_2/N}\right|^2\right)} & \text{otherwise} \end{cases}. \tag{3.12}$$

One can obtain $\check{X}$ by simply inverting the Fourier Transform. Now observe that if $\lambda \to \infty$, $\beta = 0$, and the edge approximations are exact, i.e. $\widetilde{X}_x = X_x$ and $\widetilde{X}_y = X_y$, then it follows that $\check{X} = X$. In general, selecting $\beta > 0$ will only attenuate the magnitude of any Fourier coefficients outside the set Ω. If one lets $\beta \to \infty$ (with $\lambda = \infty$, then the solution becomes equivalent to that obtained by naive Fourier back-projection, i.e. selecting $\tilde{X} = \mathscr{F}_\Omega^* \mathscr{F}_\Omega X$. This produces poor results. As a result, it is prudent to simply leave $\beta = 0$.

With the choice of $\lambda = \infty$ and $\beta = 0$, it was shown that the solution to (3.12) satisfies the following reconstruction performance guarantee [108].

Proposition 3.2. *Given approximations $\tilde{X}_x$ and $\tilde{X}_y$ of X_x and X_y, then the solution $\tilde{X}$ of Equation (3.12) will satisfy:*

$$\left\| \tilde{X} - X \right\|_2 \leq O\left(\frac{N}{\sqrt{k_1^2 + k_2^2}} \right) \max \left(\left\| \tilde{X}_x - X_x \right\|_2, \left\| \tilde{X}_y - X_y \right\|_2 \right).$$

where

$$(k_1, k_2) = \operatorname*{argmin}_{(\omega_1,\omega_2) \notin \Omega} \omega_1^2 + \omega_2^2.$$

As can be seen, the performance of this method depends on the selection of Ω. If Ω contains all the low frequencies within some radius $r = \sqrt{k_1^2 + k_2^2}$, then the final reconstruction error will be $O(N/r)$ times worse than the maximum edge reconstruction error. In general, if Ω contains mostly low frequencies, then this method will generate better results than if Ω contained mostly high frequencies. As a result, this "integration" is very appropriate in applications such as CT where Ω will consist of radial lines that congregate near the DC frequency. For the same reason, it may also be useful in MRI applications where the Fourier Space is sampled according to a spiral trajectory (see [79] and [141]). This method is referred to as GradientRec-LS in [108].

3.2.3 Numerical Examples

Figure 3.5 shows the reconstruction of a 512×512 Shepp-Logan phantom image using naive Fourier back-projection (i.e. selecting $\tilde{X} = \mathscr{F}_\Omega^* \mathscr{F}_\Omega X$), the L1TV method [79] and GradientRec-LS [108]. Only 5% of its Fourier coefficients were used. The Fourier coefficients were restricted to a radial sampling pattern as shown in Fig. 3.5(a). Figure 3.5(b) shows the result of classical Fourier back-projection which gives a relative error equal to 0.4953. The reconstruction from a TV-minimization method using the L1TV method [79] is shown in Fig. 3.5(c), which gives a relative error equal to 0.2346. Figure 3.5(d) shows the GradientRec-LS reconstruction. The recovery is near perfect with relative errors obeying 1.59×10^{-5}.

The reason why TV-minimization fails to recover the image perfectly is the following. As can be seen from Fig. 3.4 that $\|X_x\|_0 = 2972, \|X_y\|_0 = 2132$ and $\|TV(X)\|_0 = 4386$. As discussed in the previous chapter that, for a good reconstruction, the number of Fourier samples (compressive measurements) should be about

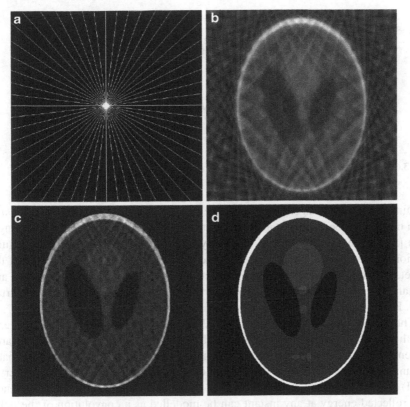

Fig. 3.5 512×512 Shepp-Logan Phantom example. (a) Fourier domain sampling pattern. (b) Back-projection. (c) L1TV reconstruction. (d) GradientRec-LS reconstruction

three to five times the number of sparse coefficients [20], [79]. This means that GradientRec-LS can recover gradients perfectly from $13107 = 0.05 \times 512 \times 512$ compressive measurements, which is approximately $6.15 \times \|X_y\|_0$ and $4.4 \times \|X_x\|_0$. Whereas for a good reconstruction, TV-minimization requires about $4 \times 4386 = 17544$ measurements. Hence, 13107 measurements are not enough for the TV-minimization to recover the underlying sparse gradient image.

3.3 Compressive Synthetic Aperture Radar Imaging

Synthetic aperture radar is a radar imaging technology that is capable of producing high resolution images of the stationary surface targets and terrain. There are four common modes of SAR: scan, stripmap, spotlight and inverse SAR (ISAR). In this chapter, we will mainly focus on the spotlight mode SAR and ISAR. In spotlight mode SAR, the radar sensor steers its antenna beam to continuously illuminate the

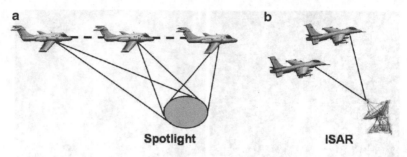

Fig. 3.6 Different modes of SAR. (a) Spotlight (b) ISAR

terrain patch being imaged. It can provide higher resolution than the stripmap and scan mode SAR because it maps a scene at multiple viewing angles during a single pass [30]. In ISAR, the radar is stationary and the target is moving. The angular motion of the target with respect to the radar can be used to form an image of the target. Differential Doppler shifts of adjacent scatters on a target are observed and the target's reflectivity function is obtained through the Doppler frequency spectrum [39]. Figure 3.6 illustrates these two SAR imaging modes.

The main advantages of SAR are its ability to operate at night and in adverse weather conditions, hence overcoming limitations of both optical and infrared systems. The basic idea of SAR is as follows: as the radar moves along its path, it transmits pulses at microwave frequencies at an uniform pulse repetition interval (PRI) which is defined as 1/PRF, where PRF is the pulse repetition frequency. The reflected energy at any instant can be modelled as a convolution of the pulse waveform with the ground reflectivity function [44, 136]. Each received pulse is pre-processed and passed on to an image formation processor. The image formation processor produces an image that is a two dimensional mapping of the illuminated scene [30]. Figure 3.7 illustrates the image formation process in spotlight mode SAR.

The two dimensional image formed is interpreted in the dimensions of range and cross-range or azimuth. The range is the direction of signal propagation and the cross-range is the direction parallel to the flight path. Sometimes the range and the cross-range samples are referred to as the fast-time and the slow-time samples, respectively. The range resolution of a SAR image is directly related to the bandwidth of the transmitted signal and the cross-range is inversely proportional to the length of the antenna aperture. Therefore, high range resolution is achieved by transmitting wide bandwidth waveforms, and high cross-range resolution is achieved by coherently processing returns transmitted from a variety of positions along a flight path to emulate a large aperture.

The standard methods for obtaining SAR images are basically based on using interpolation and the Fourier transform. One such method is known as the Polar Format Algorithm (PFA). In spotlight-mode SAR a collection of phase histories defines a set of samples in the Fourier space (k-space) of the scene on a polar

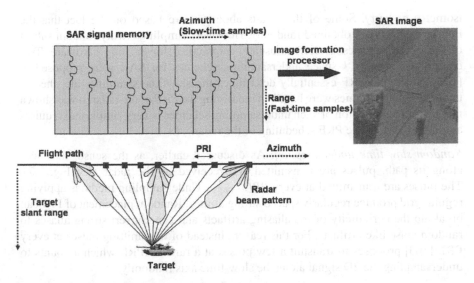

Fig. 3.7 Image formation process in spotlight mode SAR

wedge. The PFA obtains the SAR image by appropriately interpolating these polar data to a Cartesian grid and taking a two dimensional inverse Fourier transform [30]. Another popular SAR image reconstruction method, based on the tomographic formulation of SAR [92], is the filtered backprojection (FBP) method. Range migration and chirp scaling algorithms are also used for the spotlight mode SAR image reconstruction [30].

Since a SAR image is a map of the spatial distribution of the reflectivity function of stationary targets and terrain, many SAR images can be sparse or compressible in some representation such as those from wavelet or complex wavelet transform. The theory of CS allows one to recover such sparse images from a small number of random measurements provided that the undersampling results in noise like artifacts in the transform domain and an appropriate nonlinear recovery scheme is used [47], [23]. Motivated by this, compressive sampling schemes for SAR were proposed in [103], [102]. It was shown that if the SAR image is assumed to be sparse in some transform domain, then one can reconstruct a good estimate of the reflectivity profile using this new image formation algorithm that relies on using a far fewer number of waveforms than the conventional systems do and requires no changes to a radar system hardware to work.

3.3.1 Slow-time Undersampling

Design of a CS undersampling scheme for SAR entails the selection of phase histories such that the resulting sensing matrix results with a good restricted isometry property. Some of the results about CS are based on the fact that the Fourier samples are obtained randomly. However, sampling a truly random subset of the phase histories in SAR is usually impractical for existing hardware. Two compressed sensing k-space undersampling schemes for SAR were proposed in [103]. Since, the PRF essentially determines the slowtime-sampling rate, the CS undersampling schemes were based on modifying the PRF of the radar. It was shown that the implementation of such undersampling schemes is very simple and requires a minor change to the PRF scheduling of the radar.

Random slow-time undersampling: As discussed earlier, as the sensor advances along its path, pulses are transmitted and received by the radar (see Fig. 3.7). The pulses are transmitted at every $PRI = \frac{1}{PRF}$. Undersampling methods applying regular grid produce regularly spaced strong aliases. Random placement of PRI can break up the periodicity of the aliasing artifacts and can convert strong aliases to random noise like artifacts. For this reason, instead of transmitting pulses at every PRI, [103] proposes to transmit a few pulses at a random PRI, which amounts to undersampling the 2D signal along the slow-time axis randomly.

Jittered slow-time undersampling: Jittered undersampling is based on a regular undersampling which is perturbed slightly by random noise. When jittered sampling is applied the following is observed: high frequencies are attenuated, the energy lost to the attenuation appears as uniform noise, and the basic structure of the spectrum does not change. In fact, it has been shown that additive random jitter can eliminate aliasing completely [129]. Inspired by the properties of the jittered sampling, [103] proposes to apply jittered undersampling in slow-time. More analysis on the aliasing artifacts introduced by these undersampling schemes in terms of the point spread function can be found in [103].

3.3.2 Image Reconstruction

Let $X \in \mathbb{C}^{l \times l}$ be the reflectivity map to be imaged. Let $\mathbf{x} \in \mathbb{C}^{l^2 \times 1}$ be the lexicographically ordered reflectivity map X. Let Ω be some chosen set of frequencies of size $|\Omega| = M$, with $M \ll l^2$, and $\mathscr{F}_\Omega : \mathbb{C}^{l \times l} \to \mathbb{C}^M$ denote the partial Fourier transform operator restricted to Ω. Let $\mathbf{y}$ denote the collection of 2D phase histories and $\mathbf{y} = \mathscr{F}_\Omega \mathbf{x} \in \mathbb{C}^M$ represent the collection of phase histories restricted to Ω [103], [32]. That is, $\mathbf{y}$ represents the partial Fourier measurements of the reflectivity function $\mathbf{x}$ obtained by incorporating one of the slow-time undersampling schemes discussed above. Assume that $\mathbf{x}$ is sparse or compressible in some basis $\mathbf{B}$, so that $\mathbf{x} = \mathbf{B}\alpha$. Furthermore, assume that we are given the partial k-space noisy observations. So the compressive SAR observation model can be written as

$$\mathbf{y} = \mathscr{F}_\Omega \mathbf{x} + \eta = \mathscr{F}_\Omega \mathbf{B}\alpha + \eta,$$

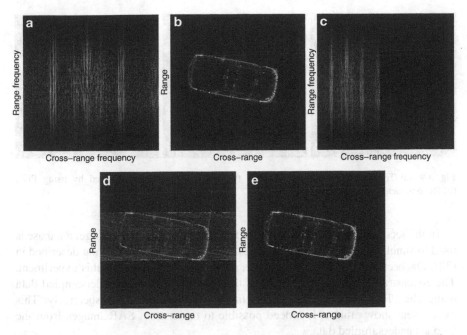

Fig. 3.8 SAAB 9000 car ISAR example. (a) Full measured data. (b) Traditional reconstruction from the full data. (c) Jittered slow-time undersampled phase histories (40% of data used). (d) Traditional reconstruction from the compressive measurements in (c). (e) Reconstructed image by solving (3.13)

where η is the measurement noise with $\|\eta\|_2 \leq \varepsilon$. Then, the reflectivity map $\mathbf{x}$ can be recovered via α by solving the following ℓ^1 minimization problem

$$\alpha_{rec} = \arg\min_{\alpha'} \|\alpha'\|_1 \text{ s. t. } \|\mathbf{y} - \mathscr{F}_\Omega \mathbf{B}\alpha'\|_2 \leq \varepsilon. \tag{3.13}$$

3.3.3 Numerical Examples

In this section, we present some simulation results that show the effectiveness of random and jittered undersampling in slow-time axis for SAR imaging. The comparison is made with the traditional PFA algorithm [30]. The SPGL1 algorithm [8] is used to solve (3.13). Figure 3.8 shows the results obtained when the jittered undersampling is applied to the ISAR data collected on a SAAB 9000 car [103]. In this experiment, only 40% of the data is used. As can be seen from Fig. 3.8, the reconstructed image from the compressed measurements, shown in Fig. 3.8(e), is identical to the one reconstructed from the full measurements, shown in Fig. 3.8(a). Figure 3.8(d) shows how the traditional reconstruction fails to recover the ISAR image from the compressive measurements shown in Fig. 3.8(c).

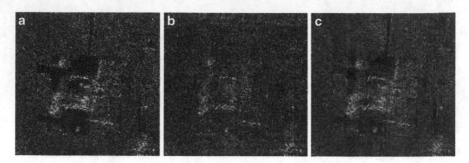

Fig. 3.9 (a) Traditional reconstruction from the full data. (b) Reconstructed by using PFA. (c) Reconstructed image by solving (3.13)

In the second example, a SAR image from the MSTAR public target database is used to simulate the SAR phase histories using the acquisition method described in [32]. Daubechies 4 wavelet is used as a sparsifying transform for this experiment. The reconstruction from only 50% of the random slow-time undersampled data using the PFA and (3.13) is shown in Fig. 3.9(b) and (c), respectively. This experiment shows that it is indeed possible to reconstruct SAR images from the k-space undersampled data.

The advantages of using such undersampling in slow-time axis in SAR is discussed in [103]. In particular, it was shown that such undersampling leads to not only reduction in data but also provides resistance to strong countermeasures, allows for imaging in wider swaths and possible reduction in antenna size.

3.4   Compressive Passive Millimeter Wave Imaging

Interest in millimeter wave (mmW) and terahertz imaging has increased in the past several years[4, 85, 162]. This interest is driven, in part, by the ability of these frequencies to penetrate poor weather and other obscurants, such as clothes and polymers. Millimeter waves are electromagnetic waves typically defined in the 30 to 300 GHz range with corresponding wavelengths between 10 to 1 mm. Radiation at these these frequencies is non-ionizing and is therefore considered safe for human use. Applications of this technology include the detection of concealed weapons, explosives and contraband [4]. Fig. 3.10 compares a visible image and corresponding 94-GHz image of two people with various weapons concealed under clothing. Note that concealed weapons are clearly detected in the mmW image.

However, when used for short range imaging (as opposed to infinite conjugate imaging in astronomy) most mmW systems have a narrow depth-of-field (DoF), the distance over which an object is considered in focus. If individuals are moving toward an imager through a corridor, the weapons would be visible only for the brief moment when they were in the depth-of-field. This is one reason individuals are scanned in portals. However, extensions to scanning over a volume could provide

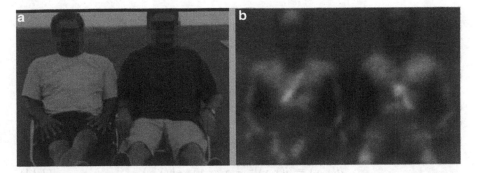

Fig. 3.10 Millimeter wave imaging through clothing. (a) Visible image of the scene. (b) Image produced using a 94-GHz imaging system

scanning without creating bottlenecks, for example, in a public marketplace where security is important but a visible display of security might be counterproductive. In [85], Mait *et al.* presented a computational imaging method to extend the depth-of-field of a passive mmW imaging system to allow for operation over a volume. The method uses a cubic phase element in the pupil plane of the system in combination with post-detection signal processing to render system operation relatively insensitive to object distance. Using this technique increased the depth-of-field of a 94 GHz imager to 68" (1727 mm), which is four times its conventional value of approximately 17" (432 mm) [85].

Unfortunately this system, as well as others discussed in the literature [4, 162], form an image by scanning a single-beam in azimuth and elevation. Although real-time mmW imaging has also been demonstrated using an array of sensors, such systems introduce complexity and are costly. Although total scan time per sample is a function of positioning and integration times, to first order at least, if one can reduce the number of samples and maintain imaging performance, one can increase scan rates with minimal detriment. To reduce the number of samples required to form an image, researchers have applied compressive sampling methods to mmW imaging [104], [5] ,[67] ,[43], [60] , [36, 98].

In this section, we describe a passive mmW imaging system with extended depth-of-field that can produce images with reduced number of samples. The method relies on using a far fewer number of measurements than the conventional systems and can reduce the sampling requirements significantly, especially when scanning over a volume [104], [105].

3.4.1 Millimeter Wave Imaging System

In [85] Mait *et al.* used a 94-GHz Stokes-vector radiometer to form images by raster scanning the system's single beam. The radiometer has a thermal sensitivity of 0.3 K with a 30-ms integration time and 1-GHz bandwidth. A Cassegrain antenna with a

24" (610 mm) diameter primary parabolic reflector and a 1.75" (44.5 mm) diameter secondary hyperbolic reflector is mounted to the front of the radiometer receiver. The position of the hyperbolic secondary is variable but fixed in our system such that the effective focal length is 6" (152.4 mm) (i.e., the system is $f/4$) and the image distance is 5.81" (147.6 mm).

One can model the 94-GHz imaging system as a linear, spatially incoherent, quasi-monochromatic system [85]. The intensity of the detected image can be represented as a convolution between the intensity of the image predicted by the geometrical optics with the system point spread function [66]

$$ii(x,y) \triangleq |i(x,y)|^2 = o_g(x,y) * *h(x,y), \tag{3.14}$$

where $**$ represents two-dimensional convolution. The function $o_g(x,y)$ represents the inverted, magnified image of the object that a ray-optics analysis of the system predicts.

The second term in Eq. (3.14), $h(x,y)$, is the incoherent point spread function (PSF) that accounts for wave propagation through the aperture

$$h(x,y) = \frac{1}{(\lambda f)^4} \left| p\left(\frac{-x}{\lambda f}, \frac{-y}{\lambda f}\right)\right|^2, \tag{3.15}$$

where $p(x/\lambda f, y/\lambda f)$ is the coherent point spread function. The function $p(x,y)$ is the inverse Fourier transform of the system pupil function $P(u,v)$,

$$p(x,y) = FT^{-1}[P(u,v)].$$

Assuming that object and image are $N \times N$ arrays, one can then rewrite Eq. (3.14) in matrix notation as

$$\mathbf{i} = \mathbf{H}\mathbf{o}_g, \tag{3.16}$$

where $\mathbf{i}$ and $\mathbf{o}_g$ are $N^2 \times 1$ lexicographically ordered column vectors representing the $N \times N$ arrays $ii(x,y)$ and $o_g(x,y)$, respectively, and $\mathbf{H}$ is the $N^2 \times N^2$ matrix that models the incoherent point spread function $h(x,y)$.

Displacement of an object from the nominal object plane of the imaging system introduces a phase error in the pupil function that increases the width of a point response and produces an out of focus image. The system's depth-of-field is defined as the distance in object space over which an object can be placed and still produce an in-focus image.

For a 94 GHz imager with an aperture diameter $D = 24$" and object distance $d_o = 180$" (4572 mm), $DoF \approx 17.4$" (442 mm) which ranges from 175.2" (4450.1 mm) to 192.6" (4892 mm).[85]

In [85], it was demonstrated how to extend the DoF using a cubic phase element in conjunction with post-detection processing. The cubic phase element $P_c(u,v)$ is

$$P_c(u,v) = \exp(j\theta_c(u,v))\texttt{rect}\left(\frac{u}{\varsigma_u}, \frac{v}{\varsigma_v}\right), \tag{3.17}$$

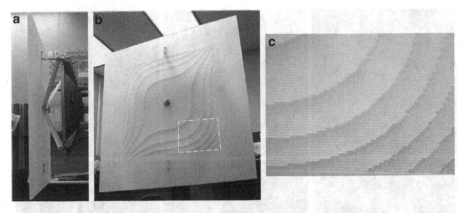

Fig. 3.11 Cubic phase element. (a) Side view of the cubic phase element mounted the antenna. (b) Front view. (c) Detail of fabricated cubic phase element

where

$$\theta_c(u,v) = (\pi\gamma)\left[\left(\frac{2u}{\xi_u}\right)^3 + \left(\frac{2v}{\xi_v}\right)^3\right]$$

and `rect` is the rectangular function. The phase function is separable in the u and v spatial frequencies and has spatial extent ξ_u and ξ_v along the respective axis. The constant γ represents the strength of the cubic phase. Fig. 3.11 shows the cubic phase element mounted on the antenna. Although fabrication introduces artifacts from spatial and phase quantization into the cubic element response, their impact on performance is negligible.[85]

Fig. 3.12 shows the measured PSFs for conventional imaging and imaging with a cubic phase. The width of the in-focus PSF at 180" (4572 mm) is approximately 2 mm, which is consistent with a 1 mm pixel width. The out-of-focus planes are at approximately four times and twice the DOF at 113" (2870 mm) and 146.5" (3721 mm), respectively, which correspond to 0.32 and 0.16 wavelengths of defocus. Note that the response of the cubic phase system is relatively unchanged, whereas the response of the conventional system changes considerably. A post-detection signal processing step is necessary to produce a well-defined sharp response [16,31,56].

If one assumes a linear post-detection process

$$i_p(x,y) = ii(x,y) * *w(x,y),\tag{3.18}$$

one can implement $w(x,y)$ as a Wiener filter in frequency space,

$$W(u,v) = \frac{H_c^*(u,v)}{|H_c(u,v)|^2 + \frac{K^{-2}\hat{\Phi}_N(u,v)}{\Phi_L(u,v)}},\tag{3.19}$$

where $H_c(u,v)$ is the optical transfer function associated with the cubic phase element, $W(u,v)$ is the Fourier transform of $w(x,y)$, the parameter K is a measure

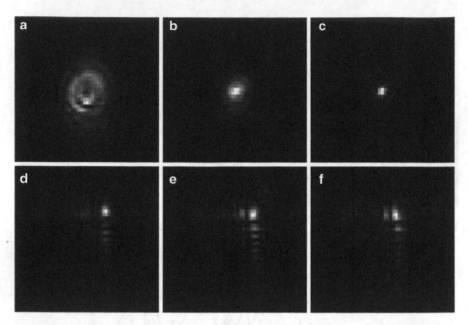

Fig. 3.12 Measured point spread functions for conventional imaging and imaging with a cubic phase. PSFs for conventional system at (a) 113" (2870 mm), (b) 146.5" (3721 mm), and (c) 180" (4572 mm). (d)-(f) PSFs for a system with cubic phase at the same distances for (a)-(c)

of the signal-to-noise ratio, and the functions $\hat{\Phi}_L$ and $\hat{\Phi}_N$ are the expected power spectra of the object and noise, respectively. The optical transfer function is usually estimated from the experimentally measured point responses. One can view the estimated $i_p(x,y)$ as a diffraction limited response. Eq. (3.18) can be written in matrix notation as

$$\mathbf{i}_p = \mathbf{Wi}$$
$$= \mathbf{WHo}_g, \tag{3.20}$$

where $\mathbf{i}_p$ is the $N^2 \times 1$ column vector corresponding to array $i_p(x,y)$ and $\mathbf{W}$ is the $N^2 \times N^2$ convolution matrix corresponding to the Wiener filter $w(x,y)$. The issue we address in the remainder of this chapter is with what fidelity can one estimate $\mathbf{i}_p$ using less than N^2 measurements of $\mathbf{i}$.

3.4.2 Accelerated Imaging with Extended Depth-of-Field

Since the objective is to form mmW images with reduced number of samples, the following sampling strategy was proposed in [104]. Because the sensor is a single-beam system that produces images by scanning in azimuth and elevation, one can reduce the number of samples by randomly undersampling in both azimuth and elevation. Mathematically, this amounts to introducing a mask in Eq. (3.14),

$$\mathbf{i}_M = \mathbf{M}\mathbf{i}$$
$$= \mathbf{M}\mathbf{H}\mathbf{o}_g, \tag{3.21}$$

where $\mathbf{i}_M$ is an $N^2 \times 1$ lexicographically ordered column vector of observations with missing information. Here, $\mathbf{M}$ is a degradation operator that removes p samples from the signal. One can construct the $N^2 \times N^2$ matrix $\mathbf{M}$ by replacing p elements in the diagonal of the $N^2 \times N^2$ identity matrix with zeros. The locations of the zeros determines which image samples are discarded.

To account for the Wiener filter in Eq. (3.18) used to process the intermediate image produced by the cubic phase element, one can write the observation model as

$$\mathbf{i}_o = \mathbf{W}\mathbf{i}_M$$
$$= \mathbf{W}\mathbf{M}\mathbf{i}$$
$$= \mathbf{W}\mathbf{M}\mathbf{H}\mathbf{o}_g. \tag{3.22}$$

One can use the relation in Eq. (3.20) to write $\mathbf{H}\mathbf{o}_g$ in terms of the diffraction limited response, $\mathbf{i}_p$,

$$\mathbf{H}\mathbf{o}_g = \mathbf{G}\mathbf{i}_p, \tag{3.23}$$

where $\mathbf{G}$ is the regularized inverse filter that corresponds to $\mathbf{W}$. With this, and assuming the presence of additive noise η, one can rewrite the observation model Eq. (3.22) as

$$\mathbf{i}_o = \mathbf{W}\mathbf{M}\mathbf{G}\mathbf{i}_p + \eta, \tag{3.24}$$

where η denotes the $N^2 \times 1$ column vector corresponding to noise, η. It is assumed that $\|\eta\|^2 = \varepsilon^2$.

Having observed $\mathbf{i}_o$ and knowing the matrices $\mathbf{W}, \mathbf{M}$ and $\mathbf{G}$, the general problem is to estimate the diffraction limited response, $\mathbf{i}_p$. Assume that $\mathbf{i}_p$ is sparse or compressible in a basis or frame Ψ so that $\mathbf{i}_p = \Psi\alpha$ with $\|\alpha\|_0 = K \ll N^2$. The observation model Eq. (3.24) can now be rewritten as

$$\mathbf{i}_o = \mathbf{W}\mathbf{M}\mathbf{G}\Psi\alpha + \eta. \tag{3.25}$$

This is a classic inverse problem whose solution can be obtained by solving the following optimization problem

$$\hat{\alpha} = \arg\min_{\alpha} \| \alpha \|_1 \text{ subject to } \|\mathbf{i}_o - \mathbf{W}\mathbf{M}\mathbf{G}\Psi\alpha\|_2 \leq \varepsilon. \tag{3.26}$$

One can clearly see the similarity between this problem and the compressed sensing problem discussed in the previous section. Once the representation vector α is

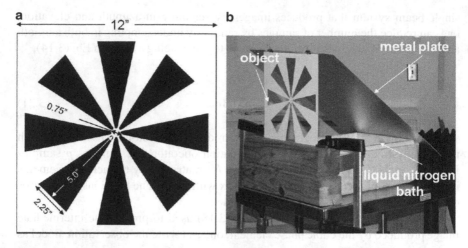

Fig. 3.13 (a) Representation of the extended object used to compare conventional and cubic-phase imaging. (b) Schematic of object illumination

estimated, one can obtain the final estimate of $\mathbf{i}_p$ as $\hat{\mathbf{i}}_p = \Psi\hat{\alpha}$. Note that the recovery of α from Eq. (3.25) depends on certain conditions on the sensing matrix $\mathbf{WMG}\Psi$ and the sparsity of α [57] as discussed in the previous section.

3.4.3   Experimental Results

In this section, we highlight some of the results presented in [104]. The extended object used in the experiments is represented in Fig. 3.13(a). Images of an extended object for conventional imaging system at 113", 146" and 180" are shown in Fig. 3.14(a)-(c), respectively. Each image is represented by 41×51 measurements, or pixels. The object size within the image is a function of optical magnification. Note that the conventional imaging system produces images with significant blurring. In contrast, even without signal processing, the images produced with cubic phase element retain more discernable characteristics of the object than the images from the conventional system, as shown in Fig. 3.14(d)-(f). It can be seen from Fig. 3.14(g)-(i) that post processing compensates for the effect of the cubic phase element and retains frequency content that is otherwise lost in a conventional system. The wider bandwidth, in addition to the noise suppressing characteristics of the Weiner filter, produce images that appear sharper than those produced by a conventional imaging system. Hence, one can extend the region over which the system generates diffraction limited images. In fact, in [85], it was shown that the *DoF* of a conventional 94-GHz imaging system can be extended from 17.4" to more than 68".

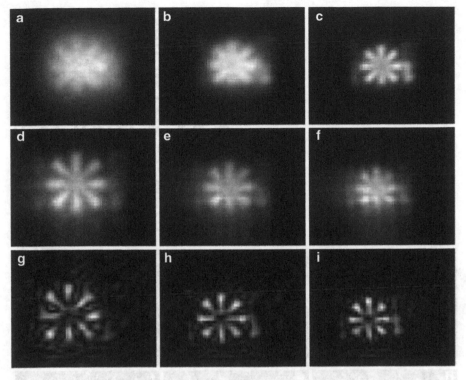

Fig. 3.14 Images from a conventional imaging system at (a) 113", (b) 146" and (c) 180". (d)-(f) Images from a system with cubic phase at the same object distances as for (a)-(c). (g)-(i) Processed images from a system with cubic phase at the same object distances as for (a)-(c)

In the first experiment, only 50% of the measured data was used. The samples were discarded according to a random undersampling pattern shown in Fig. 3.15. Figs. 3.16(a)-(c) show the sparsely sampled cubic phase data. The reconstructed images obtained by solving problem (3.26) are shown in Fig. 3.16(d)-(f). The reconstructions of the extended object are comparable to the processed images from a system with cubic phase. This can be seen by comparing Fig. 3.14(g)-(i) with Fig. 3.16(d)-(f).

3.5 Compressive Light Transport Sensing

Simulation of the appearance of real-world scenes is one of the main problems in computer graphics. To deal with this problems, image-based acquisition and relighting are often used to replace the traditional modeling and rendering of complex scenes. As a result, one has to deal with the issues related to storing and capturing the massive amount of data. For instance, consider the image shown in

Fig. 3.15 A random
undersampling pattern

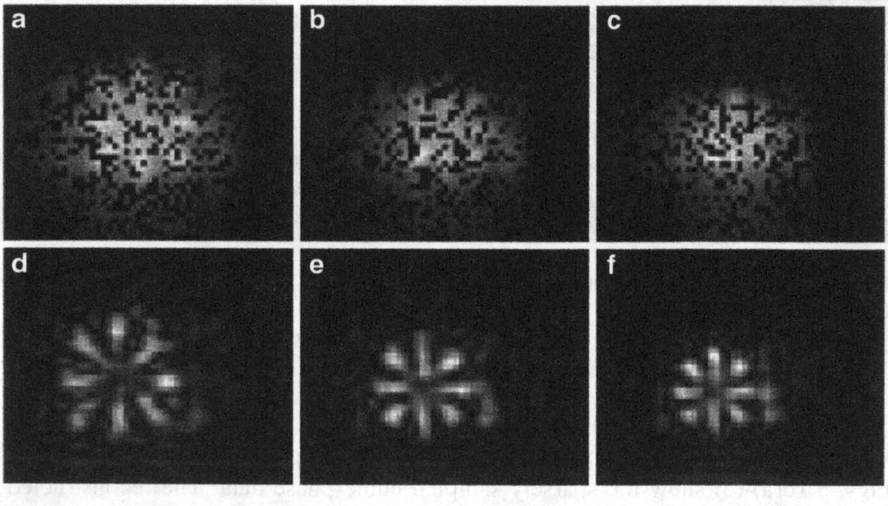

Fig. 3.16 50% Sparse sampled data from a modified imaging system at (a) 113", (b) 146", and (c) 180". (d)-(f) Diffraction limited images recovered by solving (3.26) at the same object distances as for (a)-(c)

Fig. 3.17. Acquiring high resolution datasets for this image may require thousands of photographs and gigabytes of storage [110]. To address this issue, a framework for capturing light transport data of a real scene, using the theory of compressive sensing was proposed in [110].

Image-based relighting can be written in a matrix form as follows [110]

$$\mathbf{c} = \mathbf{Tl},$$

where $\mathbf{T}$ is a $p \times n$ matrix that defines the light transport between n light sources and p camera pixels, $\mathbf{c}$ is a p dimensional vector representing pixels in the observed image, and $\mathbf{l}$ is an n dimensional vector representing illumination conditions.

Fig. 3.17 An example of relit images of a scene generated from a reflectance field captured using only 1000 nonadaptive illumination patterns [110]. (a) The scene relit with a high-frequency illumination condition. (b) The scene relit under a natural illumination condition. (c) A ground truth photograph of the scene

If we denote multiple illumination conditions as a matrix $\mathbf{L} = [\mathbf{l}_0, \cdots, \mathbf{l}_m]$ and the corresponding observations as a matrix $\mathbf{C} = [\mathbf{c}_0, \cdots, \mathbf{c}_m]$ then the acquisition stage can be compactly written as

$$\mathbf{C} = \mathbf{TL}.$$

The observations at the ith pixel are given by the following equation

$$\mathbf{c}_i = \mathbf{t}_i \mathbf{L}, \tag{3.27}$$

where $\mathbf{t}_i$ is a row of the transport matrix $\mathbf{T}$. From (3.27), one can clearly see the connection to CS where $\mathbf{L}$ plays the role of sensing matrix.

In order to apply the theory of CS to light transport data, one has to find a basis in which $\mathbf{t}_i$ are sparse or compressible. It has been shown that certain types of reflectance functions can be sparse or compressible in spherical harmonics or wavelet basis. Assuming that we have such an orthogonal basis, $\mathbf{B}$, one can write the observation matrix as

$$\mathbf{C} = \mathbf{TL}$$

$$= \mathbf{T}(\mathbf{BB}^T)\mathbf{L}$$

$$= \hat{\mathbf{T}}\mathbf{B}^T\mathbf{L}, \tag{3.28}$$

where $\hat{\mathbf{T}} = \mathbf{TB}$. Using one of the CS measurement ensembles ϕ, one can define the illumination patterns as $\mathbf{L} = \mathbf{B}\phi$. Combining this with (3.28) we obtain

$$\mathbf{C} = \hat{\mathbf{T}}(\mathbf{B}^T\mathbf{B})\phi \tag{3.29}$$

$$= \hat{\mathbf{T}}\phi. \tag{3.30}$$

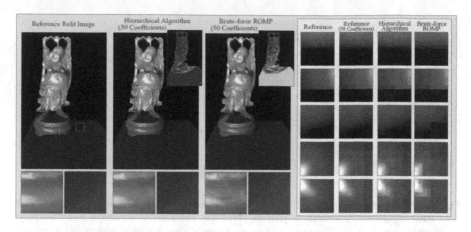

Fig. 3.18 Left: A relit image computed from a reference synthetic reflectance field. Middle: A relit image from a reflectance field inferred using the hierarchical algorithm from 512 noiseless measurements under a Gaussian ensemble. Right: A relight image from a reflectance field computed using the ROMP algorithm . Details are given at the bottom of each of the relit images. Far right: A comparison of five selected reflectance functions [110]

Each row in **C** represents a vector of compressive measurements that can be used to reconstruct the original reflectance function for that pixel via **B**. A variant of OMP algorithm known as Regularized Orthogonal Matching Pursuit (ROMP) [94] as well as a hierarchical algorithm that utilizes the coherency over the columns of **T** were used to solve the above problem [110].

Figure 3.18 shows a scene containing the Buddha model with a glossy material on a diffuse underground. The image recovered using the hierarchical algorithm from only 512 noiseless measurements with a Gaussian ensemble is shown in the middle image. The right image is the one that is obtained by using ROMP from the same measurements. As can be seen from this figure, all three images look very similar. On the right of Fig. 3.18 are the reflectance functions. See [110] for more details on the hierarchical algorithm and experimental evaluations.

Chapter 4
Compressive Sensing for Vision

In this chapter, we present an overview of some of the recent works in computer vision and image understanding that make the use of compressive sampling and sparse representation. In particular, we show have CS and sparse representation have been used for various tracking algorithms. We then present an overview of different types of compressive video cameras. Finally, we show how sparse representation framework can lead to better reconstruction of images and surfaces from gradients.

4.1 Compressive Target Tracking

In image understanding and computer vision problems such as background subtraction and target tracking one is often faced with the problem of data deluge [7]. Large amounts of data can be of detriment to tracking. Background subtraction techniques may require complicated density estimates for each pixel, which become burdensome in the presence of high-resolution imagery. Likewise, higher data dimensionality is of detriment to mean shift tracking, specifically during the required density estimation and mode search. This extra data could be due to higher sensor resolution or perhaps the presence of multiple sensors [125][126]. Therefore, new tracking strategies must be developed. The hope for finding such strategies comes from the fact that there is a substantial difference in the amount of data collected by these systems compared to the quantity of information that is ultimately of use.

Compressive sensing can help alleviate some of the challenges associated with performing classical tracking in the presence of overwhelming amounts of data. By replacing traditional cameras with compressive sensors or by making use of CS techniques in the processing of data, the actual amount of data that the system must handle can be drastically reduced. However, this capability should not come at the cost of a significant decrease in tracking performance. In this section, we will present a few methods for performing various tracking tasks that take advantage

V.M. Patel and R. Chellappa, *Sparse Representations and Compressive Sensing for Imaging and Vision*, SpringerBriefs in Electrical and Computer Engineering, DOI 10.1007/978-1-4614-6381-8_4, © The Author(s) 2013

of CS in order to reduce the quantity of data that must be processed. Specifically, recent methods using CS to perform background subtraction, more general signal tracking, multi-view visual tracking, and particle filtering will be discussed.

4.1.1 Compressive Sensing for Background Subtraction

One of the most intuitive applications of compressive sensing in visual tracking is the modification of background subtraction such that it is able to operate on compressive measurements. Background subtraction aims to segment the object-containing foreground from the uninteresting background. This process not only helps to localize objects, but also reduces the amount of data that must be processed at later stages of tracking. However, traditional background subtraction techniques require that the full image be available before the process can begin. Such a scenario is reminiscent of the problem that CS aims to address. Noting that the foreground signal (image) is sparse in the spatial domain, [33] have presented a technique via which background subtraction can be performed on compressive measurements of a scene, resulting in a reduced data rate while simultaneously retaining the ability to reconstruct the foreground. More recently, a modification which adaptively adjusts the number of collected compressive measurements based on the dynamic foreground sparsity typical to surveillance data has been proposed in [152].

Denote the images comprising a video sequence as $\{\mathbf{x}_t\}_{t=0}^{\infty}$, where $\mathbf{x}_t \in \mathbb{R}^N$ is the vectorized image captured at time t. Cevher *et al.* [33] model each image as the sum of foreground and background components $\mathbf{f}_t$ and $\mathbf{b}_t$, respectively. That is,

$$\mathbf{x}_t = \mathbf{f}_t + \mathbf{b}_t. \tag{4.1}$$

Assume $\mathbf{x}_t$ is sensed using $\boldsymbol{\Phi} \in \mathbb{C}^{M \times N}$ to obtain compressive measurements $\mathbf{y}_t = \boldsymbol{\Phi}\mathbf{x}_t$. If $\Delta(\boldsymbol{\Phi},\mathbf{y})$ represents a CS decoding procedure, then the proposed method for estimating $\mathbf{f}_t$ from $\mathbf{y}_t$ is

$$\hat{\mathbf{f}}_t = \Delta(\boldsymbol{\Phi}, \mathbf{y} - \mathbf{y}_t^b), \tag{4.2}$$

where it is assumed that $\mathbf{y}_t^b = \boldsymbol{\Phi}\mathbf{b_t}$ is known via an estimation and update procedure.

To begin, $\mathbf{y}_0^b$ is initialized using a sequence of N compressively sensed background-only frames $\{\mathbf{y}_j^b\}_{j=1}^N$ that appear before the sequence of interest begins. These measurements are assumed to be realizations of a multivariate Gaussian random variable, and the maximum likelihood (ML) procedure is used to estimate its mean as $\mathbf{y}_0^b = \frac{1}{N}\sum_{j=1}^N \mathbf{y}_j^b$. This estimate is used as the known background for $t = 0$ in (4.2). Since the background typically changes over time, a method is proposed for updating the background estimate based on previous observations. Specifically, the following is proposed:

$$\mathbf{y}_{t+1}^b = \alpha(\mathbf{y}_t - \boldsymbol{\Phi}\Delta(\boldsymbol{\Phi}, \mathbf{y}_{t+1}^{ma})) + (1-\alpha)\mathbf{y}_t^b \tag{4.3}$$

$$\mathbf{y}_{t+1}^{ma} = \gamma\mathbf{y}_t + (1-\gamma)\mathbf{y}_t^{ma}, \tag{4.4}$$

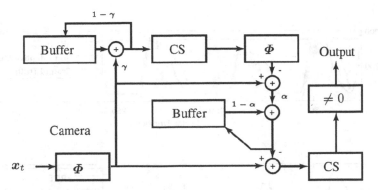

Fig. 4.1 Block diagram of the compressive sensing for background subtraction technique. Figure originally appears in [33]

where $\alpha, \gamma \in (0,1)$ are learning rate parameters and $\mathbf{y}_{t+1}^{ma}$ is a moving average term. This method compensates for both gradual and sudden changes to the background. A block diagram of the proposed system is shown in Fig. 4.1.

The above procedure assumes a fixed $\Phi \in \mathbb{C}^{M \times N}$. Therefore, M compressive measurements of $\mathbf{x}_t$ are collected at time t regardless of its content. It is not hard to imagine that the number of significant components of $\mathbf{f}_t$, k_t, might vary widely with t. For example, consider a scenario in which the foreground consists of a single object at $t = t_0$, but many more at $t = t_1$. Then $k_1 > k_0$, and $M > Ck_1 \log N$ implies that $\mathbf{x}_{t_0}$ has been oversampled due to the fact that only $M > Ck_0 \log N$ measurements are necessary to obtain a good approximation of $\mathbf{f}_{t_0}$. Foregoing the ability to update the background, a modification to the above method for which the number of compressive measurements at each frame, M_t, can vary has been proposed in [152].

Such a scheme requires a different measurement matrix for each time instant, i.e. $\Phi_t \in \mathbb{C}^{M_t \times N}$. To form Φ_t, one first constructs $\Phi \in \mathbb{C}^{N \times N}$ via standard CS measurement matrix construction techniques. Φ_t is then formed by selecting only the first M_t rows of Φ and column-normalizing the result. The fixed background estimate, $\mathbf{y}^b$, is estimated from a set of measurements of the background only obtained via Φ. In order to use this estimate at each time instant t, $\mathbf{y}_t^b$ is formed by retaining only the first M_t components of $\mathbf{y}^b$.

In parallel to Φ_t, the method also requires an extra set of compressive measurements via which the quality of the foreground estimate, $\hat{\mathbf{f}}_t = \Delta(\Phi_t, \mathbf{y}_t - \mathbf{y}_t^b)$, is determined. These are obtained via a *cross validation* matrix $\Psi \in \mathbb{C}^{r \times N}$, which is constructed in a manner similar to Φ. r depends on the desired accuracy of the cross validation error estimate (given below), is negligible compared to N, and constant for all t. In order to use the measurements $\mathbf{z}_t = \Psi \mathbf{x}_t$, it is necessary to perform background subtraction in this domain via an estimate of the background, $\mathbf{z}^b$, which is obtained in a manner similar to $\mathbf{y}^b$ above.

The quality of $\hat{\mathbf{f}}_t$ depends on the relationship between k_t and M_t. Using a technique operationally similar to cross validation, an estimate of $\|\mathbf{f}_t - \hat{\mathbf{f}}_t\|_2$, i.e., the

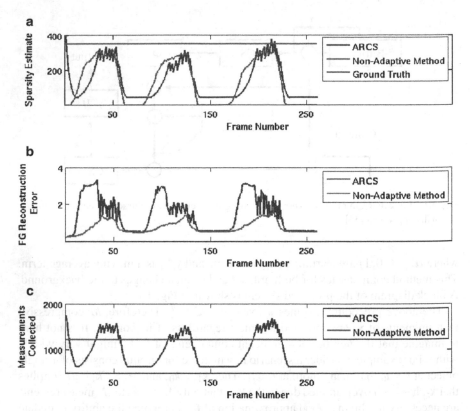

Fig. 4.2 Comparison between ARCS and a non-adaptive method for a dataset consisting of vehicles moving in and out of the field of view. (a) Foreground sparsity estimates for each frame, including ground truth. (b) ℓ_2 foreground reconstruction error. (c) Number of measurements required. Note the measurements savings provided by ARCS for most frames, and its ability to track the dynamic foreground sparsity. Figure originally appears in [152]

error between the true foreground and the reconstruction provided by Δ at time t, is provided by $\|(\mathbf{z}_t - \mathbf{z}^b) - \Psi \hat{\mathbf{f}}_t\|_2$. M_{t+1} is set to be greater or less than M_t depending on the hypothesis test

$$\|(\mathbf{z}_t - \mathbf{z}^b) - \Psi \hat{\mathbf{f}}_t\|_2 \lessgtr \tau_t. \qquad (4.5)$$

Here, τ_t is a quantity set based on the expected value of $\|\mathbf{f}_t - \hat{\mathbf{f}}_t\|_2$ assuming M_t to be large enough compared to k_t. The overall algorithm is termed *adaptive rate compressive sensing (ARCS)*, and the performance of this method compared to a non-adaptive approach is shown in Fig. 4.2.

Both techniques assume that the tracking system can only collect compressive measurements and provide a method by which foreground images can be reconstructed. These foreground images can then be used just as in classical tracking applications. Thus, CS has provided a means by which to reduce the up-front data costs associated with the system while retaining the information necessary to track.

4.1.2 Kalman Filtered Compressive Sensing

A more general problem regarding signal tracking using compressive observations is considered in [146]. The signal being tracked, $\{\mathbf{x}_t\}_{t=0}^{\infty}$, is assumed to be both sparse and have a slowly-changing sparsity pattern. Given these assumptions, if the support set of $\mathbf{x}_t$, T_t, is known, the relationship between $\mathbf{x}_t$ and $\mathbf{y}_t$ can be written as:

$$\mathbf{y}_t = \Phi_{T_t}(\mathbf{x})_{T_t} + \mathbf{w}_t. \tag{4.6}$$

Above, Φ is the CS measurement matrix, and Φ_{T_t} retains only those columns of Φ whose indices lie in T_t. Likewise, $(\mathbf{x}_t)_{T_t}$ contains only those components corresponding to T_t. Finally, $\mathbf{w}_t$ is assumed to be zero mean Gaussian noise. If $\mathbf{x}_t$ is assumed to also follow the state model $\mathbf{x}_t = \mathbf{x}_{t-1} + \mathbf{v}_t$ with $\mathbf{v}_t$ zero mean Gaussian noise, then the MMSE estimate of $\mathbf{x}_t$ from $\mathbf{y}_t$ can be computed using a Kalman filter instead of a CS decoder.

The above is only valid if T_t is known, which is often not the case. This is handled by using the Kalman filter output to detect changes in T_t and re-estimate it if necessary. $\tilde{\mathbf{y}}_{t,f} = \mathbf{y}_t - \Phi\hat{\mathbf{x}}$, the filter error, is used to detect changes in the signal support via a likelihood ratio test given by

$$\tilde{\mathbf{y}}_{t,f}' \Sigma \tilde{\mathbf{y}}_{t,f} \gtrless \tau \tag{4.7}$$

where τ is a threshold and Σ is the filtering error covariance. If the term on the left hand side exceeds the threshold, then changes to the support set are found by applying a procedure based on the Dantzig selector. Once T_t has been re-estimated, $\hat{\mathbf{x}}$ is re-evaluated using this new support set.

The above algorithm is useful in surveillance scenarios when objects under observation are stationary or slowly-moving. Under such assumptions, this method is able to perform signal tracking with a low data rate and low computational complexity.

4.1.3 Joint Compressive Video Coding and Analysis

Cossalter et al. [42] consider a collection of methods via which systems utilizing compressive imaging devices can perform visual tracking. Of particular note is a method referred to as *joint compressive video coding and analysis*, in which the tracker output is used to improve the overall effectiveness of the system. Instrumental to this method is work from theoretical CS literature which proposes a weighted decoding procedure that iteratively determines the locations and values of the (nonzero) sparse vector coefficients. Modifying this decoder, the joint coding and analysis method utilizes the tracker estimate to directly influence the weights. The result is a foreground estimate of higher quality compared to one obtained via standard CS decoding techniques.

The weighted CS decoding procedure calculates the foreground estimate via

$$\hat{\mathbf{f}} = \min_{\theta} \|\mathbf{W}\theta\|_1 \quad \text{s.t.} \quad \|\mathbf{y}^f - \Phi\theta\|_2 \leq \sigma, \tag{4.8}$$

where $\mathbf{y}^f = \mathbf{y} - \mathbf{y}^b$, $\mathbf{W}$ is a diagonal matrix with weights $[w(1)\dots w(N)]$, and σ captures the expected measurement and quantization noise in $\mathbf{y}^f$. Ideally, the weights are selected according to

$$w(i) = \frac{1}{|f(i)| + \varepsilon}, \tag{4.9}$$

where $f(i)$ is the value of the ith coefficient in the true foreground image. Of course, these values are not known in advance, but the closer the weights are to their actual value, the more accurate $\hat{\mathbf{f}}$ becomes. The joint coding and analysis approach utilizes the tracker output in selecting appropriate values for these weights.

The actual task of tracking is accomplished using a particle filter. The state vector for an object at time t is denoted by $\mathbf{z}_t = [\mathbf{c}_t \, \mathbf{s}_t \, \mathbf{u}_t]$, where $\mathbf{s}_t$ represents the size of the bounding box defined by the object appearance, $\mathbf{c}_t$ the centroid of this box, and $\mathbf{u}_t$ the object velocity in the image plane. A suitable kinematic motion model is utilized to describe the expected behavior of these quantities with respect to time, and foreground reconstructions are used to generate observations.

Assuming the foreground reconstruction $\hat{\mathbf{f}}_t$ obtained via decoding the compressive observations from time t is accurate, a reliable tracker estimate can be computed. This estimate, $\hat{\mathbf{z}}_t$, can then be used to select values for the weights $[w(1)\dots w(N)]$ at time $t+1$. If the weights are close to their ideal value (4.9), the value of $\hat{\mathbf{f}}_{t+1}$ obtained from the weighted decoding procedure will be of higher quality than that obtained from a more generic CS decoder. [42] explores two methods via which the weights at time $t+1$ can be selected using $\hat{\mathbf{f}}_t$ and $\hat{\mathbf{z}}_t$. The best of these consists of three steps: 1) thresholding the entries of $\hat{\mathbf{f}}_t$, 2) translating the thresholded silhouettes for a single time step according to the motion model and $\hat{\mathbf{z}}_t$, and 3) dilating the translated silhouettes using a predefined dilation element. The final step accounts for uncertainty in the change of object appearance from one frame to the next. The result is a modified foreground image, which can then be interpreted as a prediction of $\mathbf{f}_{t+1}$. This prediction is used to define the weights according to (4.9), and the weighted decoding procedure is used to obtain $\hat{\mathbf{f}}_{t+1}$.

The above method is repeated at each new time instant. For a fixed compressive measurement rate, it is shown to provide more accurate foreground reconstructions than decoders that do not take advantage of the tracker output. Accordingly, it is also the case that such a method is able to more successfully tolerate lower bit rates. These results reveal the benefit of using the high level tracker information in compressive sensing systems.

4.1.4 Compressive Sensing for Multi-View Tracking

Another direct application of CS to a data-rich tracking problem is presented by [116]. Specifically, a method for using multiple sensors to perform multi-view tracking employing a coding scheme based on compressive sensing is developed. Assuming that the observed data contains no background component (this could be realized, e.g., by preprocessing using any of the background subtraction techniques previously discussed), the method uses known information regarding the sensor geometry to facilitate a common data encoding scheme based on CS. After data from each camera is received at a central processing station, it is fused via CS decoding and the resulting image or three dimensional grid can be used for tracking.

The first case considered is one where all objects of interest exist in a known ground plane. It is assumed that the geometric transformation between it and each sensor plane is known. That is, if there are C cameras, then the *homographies* $\{\mathbf{H}_j\}_{j=1}^{C}$ are known. The relationship between coordinates (u,v) in the jth image and the corresponding ground plane coordinates (x,y) is determined by $\mathbf{H}_j$ as

$$\begin{bmatrix} u \\ v \\ 1 \end{bmatrix} \sim \mathbf{H}_j \begin{bmatrix} x \\ y \\ 1 \end{bmatrix}, \tag{4.10}$$

where the coordinates are written in accordance with their homogeneous representation. Since $\mathbf{H}_j$ can vary widely across the set of cameras due to varying viewpoint, an encoding scheme designed to achieve a common data representation is presented. First, the ground plane is sampled, yielding a discrete set of coordinates $\{(x_i,y_i)\}_{i=1}^{N}$. An occupancy vector, $\mathbf{x}$, is defined over these coordinates, where $\mathbf{x}(n) = 1$ if foreground is present at the corresponding coordinates and is 0 otherwise. For each camera's observed foreground image in the set $\{\mathbf{I}_j\}_{j=1}^{C}$, an occupancy vector $\mathbf{y}'_j$ is formed as $\mathbf{y}'_j(i) = \mathbf{I}_j(u_i,v_i)$, where (u_i,v_i) are the (rounded) image plane coordinates corresponding to (x_i,y_i) obtained via (4.10). Thus, $\mathbf{y}'_j = \mathbf{x} + \mathbf{e}_j$, where $\mathbf{e}_j$ represents any error due to the coordinate rounding and other noise. Figure 4.3 illustrates the physical configuration of the system.

Noting that $\mathbf{x}$ is often sparse, the camera data $\{\mathbf{y}'_j\}_{j=1}^{C}$ is encoded using compressive sensing. First, C measurement matrices $\{\boldsymbol{\Phi}_j\}_{j=1}^{C}$ of equal dimension are formed according to a construction that affords them the RIP of appropriate order for $\mathbf{x}$. Next, the camera data is projected into the lower-dimensional space by computing $\mathbf{y}_j = \boldsymbol{\Phi}_j \mathbf{y}'_j$, $j = 1,\ldots,C$. This lower-dimensional data is transmitted to a central station, where it is ordered into the following structure:

$$\begin{bmatrix} \mathbf{y}_1 \\ \vdots \\ \mathbf{y}_C \end{bmatrix} = \begin{bmatrix} \boldsymbol{\Phi}_1 \\ \vdots \\ \boldsymbol{\Phi}_C \end{bmatrix} \mathbf{x} + \begin{bmatrix} \mathbf{e}_1 \\ \vdots \\ \mathbf{e}_C \end{bmatrix} \tag{4.11}$$

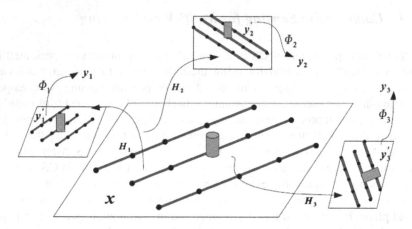

Fig. 4.3 Physical diagram capturing the assumed setup of the multi-view tracking scenario. Figure originally appears in [116]

which can be written as $\mathbf{y} = \boldsymbol{\Phi}\mathbf{x} + \mathbf{e}$. This is a noisy version of the standard CS problem, and an estimate of $\mathbf{x}$ can be found as follows

$$\hat{\mathbf{x}} = \min_{\mathbf{z} \in \mathbb{R}^N} \|\mathbf{z}\|_1 \quad \text{subject to} \quad \|\boldsymbol{\Phi}\mathbf{z} - \mathbf{y}\|_2 \leq \|\mathbf{e}\|_2. \tag{4.12}$$

The estimated occupancy grid (formed, e.g., by thresholding $\hat{\mathbf{x}}$) can then be used as input to subsequent tracker components.

The procedure mentioned above has been extended to three dimensions, where $\mathbf{x}$ represents an occupancy grid over 3D space, and the geometric relationship in (4.10) is modified to account for the added dimension. The rest of the process is entirely similar to the two dimensional case. Of particular note is the advantage in computational complexity: it is only on the order of the dimension of $\mathbf{x}$ as opposed to the number of measurements received.

4.1.5 Compressive Particle Filtering

The final application of compressive sensing in tracking presented in this chapter is the compressive particle filtering algorithm developed by [150]. As in Sect. 4.1.1, it is assumed that the system uses a sensor that is able to collect compressive measurements. The goal is to obtain tracks *without* having to perform CS decoding. That is, the method solves the sequential estimation problem using the compressive measurements directly. Specifically, the algorithm is a modification to the particle filter.

First, the system is formulated in state space, where the state vector at time t is given by

$$\mathbf{s}_t = [s_t^x \ s_t^y \ \dot{s}_t^x \ \dot{s}_t^y \ \psi_t]^T. \tag{4.13}$$

(s_t^x, s_t^y) and $(\dot{s}_t^x, \dot{s}_t^y)$ represent the object position and velocity in the image plane, and ψ_t is a parameter specifying the width of an appearance kernel. The appearance kernel is taken to be a Gaussian function defined over the image plane and centered at (s_t^x, s_t^y) with i.i.d. component variance proportional to ψ_t. That is, given $\mathbf{s}_t$, the jth component of the vectorized image, $\mathbf{z}_t$, is defined as

$$z_t^j(\mathbf{s}_t) = C_t \exp\{-\psi_t(\begin{bmatrix} s_k^x \\ s_k^y \end{bmatrix} - \mathbf{r}^j)\}, \tag{4.14}$$

where $\mathbf{r}^j$ specifies the two dimensional coordinate vector belonging to the jth component of $\mathbf{z}_t$.

The state equation is given by

$$\mathbf{s}_{t+1} = f_t(\mathbf{s}_t, \mathbf{v}_t) = \mathbf{D}\mathbf{s}_t + \mathbf{v}_t, \tag{4.15}$$

where

$$D = \begin{bmatrix} 1 & 0 & 1 & 0 & 0 \\ 0 & 1 & 0 & 1 & 0 \\ 0 & 0 & 1 & 0 & 0 \\ 0 & 0 & 0 & 1 & 0 \\ 0 & 0 & 0 & 0 & 1 \end{bmatrix} \tag{4.16}$$

and $\mathbf{v}_t \sim \mathcal{N}(\mathbf{0}, \mathrm{diag}(\alpha))$ for a preselected noise variance vector α.

The observation equation specifies the mapping from the state to the observed compressive measurements $\mathbf{y}_t$. If Φ is the CS measurement matrix used to sense $\mathbf{z}_t$, this is given by

$$\mathbf{y}_t = \Phi \mathbf{z}_t(\mathbf{s}_t) + \mathbf{w}_t, \tag{4.17}$$

where $\mathbf{w}_t$ is zero-mean Gaussian measurement noise with covariance Σ.

With the above specified, the bootstrap particle filtering algorithm can be used to sequentially estimate $\mathbf{s}_t$ from the observations $\mathbf{y}_t$. Specifically, the importance weights belonging to candidate samples $\{\tilde{\mathbf{s}}_t^{(i)}\}_{i=1}^N$ can be found via

$$\tilde{w}_t^{(i)} = p(\mathbf{y}_t | \tilde{\mathbf{s}}_t^{(i)}) = \mathcal{N}(\mathbf{y}_t; \Phi \mathbf{z}_t(\tilde{\mathbf{s}}_t^{(i)}), \Sigma) \tag{4.18}$$

and rescaling to normalize across all i. These importance weights can be calculated at each time step without having to perform CS decoding on $\mathbf{y}$. In some sense, the filter is acting purely on compressive measurements, and hence the name "compressive particle filter."

4.2 Compressive Video Processing

As we saw in the previous sections, recent advances in CS have led to the development of many imaging devices that sense static signals such as images at measurement rates lower than the Nyquist rate. Tremendous progress has also been made in compressively sensing temporal sequences or videos. In this chapter, we highlight some of these techniques for compressively sensing videos. In particular, we present an overview of compressive sensing of high-speed periodic videos [147], programable pixel camera for compressively sensing high speed videos [117] and compressive acquisition of dynamic scenes [127].

4.2.1 Compressive Sensing for High-Speed Periodic Videos

Processing of periodic signals is an important problem in many applications. For instance, several human and animal biological processes such as heart-beat, breathing, several cellular processes, industrial automation processes and everyday objects such as hand-mixer and blender all generate periodic processes. Nevertheless, we are often unaware of the inner workings of some of these high-speed processes because they occur at a far greater speed than can be perceived by the human eye. In [147], a techniques based on compressive sampling was proposed that can turn an off-the-shelf video camera into a powerful high-speed video camera for observing periodic events. Essentially, the idea is to capture periodic phenomena by coding the shutter of a low frame-rate camera during every frame and reconstruct the phenomena at a higher rate by exploiting sparsity of periodic signals.

Strobing is often used in entertainment, medical imaging and industrial applications to visualize and capture high-speed visual phenomena. In case of periodic phenomenon, strobing is commonly used to achieve aliasing and generate lower beat frequencies. While strobing performs effectively when the scene consists of a single frequency with a narrow sideband, it is difficult to visualize multiple or a wider band of frequencies simultaneously. Instead of direct observation of beat frequencies, one can exploit a computational camera approach based on different sampling sequences. The key idea is to measure appropriate linear combinations of the periodic signal and then decode the signal by exploiting the sparsity of the signal in Fourier domain. Figure 4.4 illustrates the idea behind coded strobing camera (CSC).

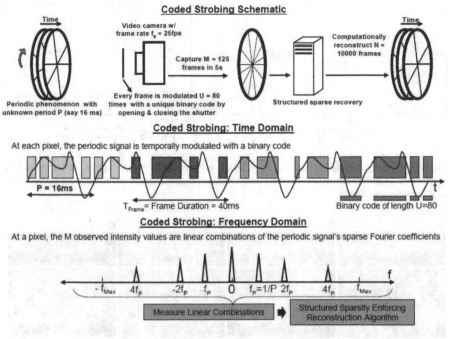

Fig. 4.4 CSC: A fast periodic visual phenomenon is recorded by a normal video camera (25 fps) by randomly opening and closing the shutter at high speed (2000 Hz). The phenomenon is accurately reconstructed from the captured frames at the high-speed shutter rate (2000 fps) [147]

Let $\mathbf{x}$ represent the luminance at a pixel. Then assuming that $\mathbf{x}$ is bandlimited, it can be represented as

$$\mathbf{x} = \mathbf{Bs}, \tag{4.19}$$

where the columns of $\mathbf{B}$ contain Fourier basis elements and since the signal $x(t)$ is assumed to be periodic the coefficients $\mathbf{s}$ are sparse in Fourier basis. The observed intensity values $\mathbf{y}$ at a given pixel can be represented as

$$\mathbf{y} = \mathbf{Cx} + \eta, \tag{4.20}$$

where the $M \times N$ matrix $\mathbf{C}$ performs both the modulation and integration for frame duration and η represents the observation noise. Using (4.19) the observation model (4.20) can be written as

$$\mathbf{y} = \mathbf{Cx} + \eta = \mathbf{CBs} + \eta = \mathbf{As} + \eta, \tag{4.21}$$

where $\mathbf{A} = \mathbf{CB}$. This observation model is illustrated in Fig. 4.5.

To reconstruct the high-speed periodic signal $\mathbf{x}$, it suffices to reconstruct its Fourier coefficients $\mathbf{s}$ from modulated intensity observations $\mathbf{y}$ of the scene. Since the system of equations (4.21) is underdetermined, in order to obtain robust

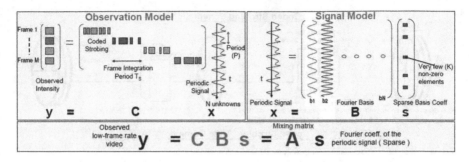

Fig. 4.5 Observation model shows the capture process of the CSC where different colors correspond to different frames and the binary shutter sequence is depicted using the presence or absence of color [147]

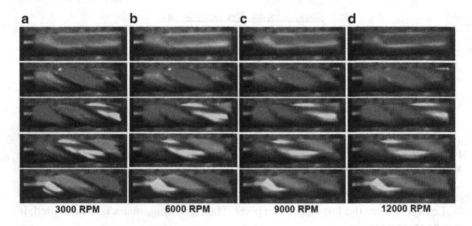

Fig. 4.6 Tool bit rotating at different rpm captured using coded strobing: Top row shows the coded images acquired by a PGR Dragonfly2 at 25 fps, with an external FLC shutter fluttering at 2000 Hz. (a)-(d) Reconstruction results, at 2000 fps, of a tool bit rotating at 3000, 6000, 9000 and 12000 rpm respectively. For better visualization, the tool was painted with color prior to the capture [147]

solutions further knowledge about the signal needs to be used. Since the Fourier coefficients **s**, of a periodic signal **x**, are sparse, a reconstruction technique enforcing sparsity of **s** can be used to recover the periodic signal **x**. This can be done bys solving the following optimization problem

$$\hat{\mathbf{s}} = \arg\min \|\mathbf{s}\|_1 \text{ subject to } \|\mathbf{y} - \mathbf{As}\|_2 \leq \varepsilon. \quad (4.22)$$

Note that the recovery of **s** by solving the above optimization problem will depend on the conditioning of **A** as well as the sparsity of **s**. See [147] for more details on the properties of **A** and various reconstruction algorithms for recovering **s** from **y**. Figure 4.6 shows the reconstruction results at 2000 fps of a tool bit rotating at 3000, 6000, 9000 and 12000 rpm, respectively.

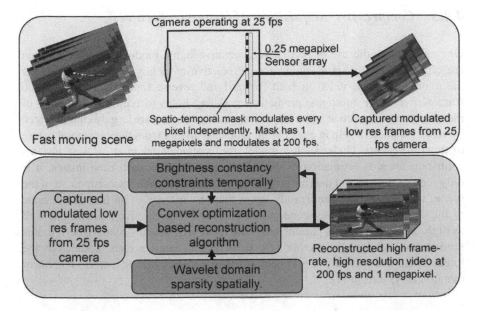

Fig. 4.7 Programmable Pixel Compressive Camera (P2C2): Each pixel of a low frame rate, low resolution camera is modulated independently with a fast, high resolution modulator (LCOS or DMD). The captured modulated low resolution frames are used with accompanying brightness constancy constraints and a wavelet domain sparsity model in a convex optimization framework to recover high resolution, high-speed video [117]

4.2.2 Programmable Pixel Compressive Camera for High Speed Imaging

In order to capture fast phenomena at frame rates higher than the camera sensor, in [117], an imaging architecture for compressive video sensing termed programmable pixel compressive camera (P2C2) was proposed. The imaging architecture of P2C2 is shown in Fig. 4.7.

P2C2 consists of a normal 25 fps, low resolution video camera, with a high resolution, high frame-rate modulating device such as a Liquid Crystal on Silicon (LCOS) or a Digital Micromirror Device (DMD) array. The modulating device modulates each pixel independently in a pre-determined random fash-ion at a rate higher than the acquisition frame rate of the camera. Thus, each observed frame at the camera is a coded linear combination of the voxels of the underlying highspeed video frames. Both low frame-rate video cameras and high frame-rate amplitude modulators (DMD/LCOS) are inexpensive and this results in significant cost reduction. Further, the capture bandwidth is significantly reduced due to P2C2s compressive imaging architecture. The underlying high resolution, high-speed frames are recovered from the captured low resolution frames by exploiting temporal redundancy in the form of brightness constancy and spatial redundancy through transform domain sparsity in a convex optimization framework [117].

4.2.3   Compressive Acquisition of Dynamic Textures

One can explore the use of predictive/generative signal models for video CS that are characterized by static parameters. Predictive modeling provides a prior for the evolution of the video in both forward and reverse time. By relating video frames over small durations, predictive modeling helps to reduce the number of measurements required at a given time instant. Models that are largely characterized by static parameters help in eliminating problems arising from the ephemeral nature of dynamic events. Under such a model, measurements taken at *all* time instants contribute towards estimation of the static parameters. At each time instant, it is only required to sense at the rate sufficient to acquire the dynamic component of the scene, which could be significantly lower than the sparsity of an individual frame of the video. One dynamic scene model that exhibits predictive modeling as well as high-dimensional static parameters is the linear dynamical system (LDS). In this section, we highlight the methods for the CS of dynamic scenes modeled as the linear dynamical system. We first give a background on dynamic textures and linear dynamical systems [127].

4.2.3.1   Dynamic Textures and Linear Dynamical Systems

Linear dynamical systems represent a class of parametric models for time-series data including dynamic textures [53], traffic scenes [34], and human activities [148], [144]. Let $\{\mathbf{y}_t, t = 0, \ldots, T\}$ be a sequence of frames indexed by time t. The LDS model parameterizes the evolution of $\mathbf{y}_t$ as follows:

$$\mathbf{y}_t = C\mathbf{x}_t + \mathbf{w}_t \quad \mathbf{w}_t \sim N(\mathbf{0}, R), R \in \mathbb{R}^{N \times N} \tag{4.23}$$

$$\mathbf{x}_{t+1} = A\mathbf{x}_t + \mathbf{v}_t \quad \mathbf{v}_t \sim N(\mathbf{0}, Q), Q \in \mathbb{R}^{d \times d} \tag{4.24}$$

where $\mathbf{x}_t \in \mathbb{R}^d$ is the hidden state vector, $A \in \mathbb{R}^{d \times d}$ the transition matrix, and $C \in \mathbb{R}^{N \times d}$ is the observation matrix.

Given the observations $\{\mathbf{y}_t\}$, the truncated SVD of the matrix $[\mathbf{y}]_{1:T} = [\mathbf{y}_1, \mathbf{y}_2, \ldots, \mathbf{y}_T]$ can be used to estimate both C and A. In particular, an estimate of the observation matrix C is obtained using the truncated SVD of $[\mathbf{y}]_{1:T}$. Note that the choice of C is unique only up to a $d \times d$ linear transformation. That is, given $[\mathbf{y}]_{1:T}$, we can define $\widehat{C} = UL$, where L is an invertible $d \times d$ matrix. This represents our choice of coordinates in the subspace defined by the columns of C. This lack of uniqueness leads to structured sparsity patterns which can be exploited in the inference algorithms.

4.2.3.2 Compressive Acquisition of LDS

In the rest of the chapter, we use the following notation. At time t, the image observation (the t-th frame of the video) is $\mathbf{y}_t \in \mathbb{R}^N$ and the hidden state is $\mathbf{x}_t \in \mathbb{R}^d$ such that $\mathbf{y}_t = C\mathbf{x}_t$, where $C \in \mathbb{R}^{N \times d}$ is the observation matrix. We use $\mathbf{z}$ to denote compressive measurements and Φ and Ψ to denote the measurement and sparsifying matrices respectively. We use ":" subscripts to denote sequences, such as $\mathbf{x}_{1:T} = \{\mathbf{x}_1, \mathbf{x}_2, \dots, \mathbf{x}_T\}$ and $[\cdot]_{1:T}$ to denote matrices, such as $[\mathbf{y}]_{1:T}$ is the $N \times T$ matrix formed by $\mathbf{y}_{1:T}$ such that the k-th column is $\mathbf{y}_k$.

One of the key features of an LDS is that the observations $\mathbf{y}_t$ lie in the subspace spanned by the columns of the matrix C. The subspace spanned by C forms a static parameter of the system. Estimating C and the dynamics encoded in the state sequence $\mathbf{x}_{1:T}$ is sufficient for reconstructing the video. For most LDSs, $N \gg d$, thereby making C much higher dimensional than the state sequence $\{\mathbf{x}_t\}$. In this sense, the LDS models the video using high information rate static parameters (such as C) and low information rate dynamic components (such as $\mathbf{x}_t$). This relates to our initial motivation for identifying signal models with parameters that are largely static. The subspace spanned by C is static, and hence, we can "pool" measurements over time to recover C.

Further, given that the observations $\mathbf{y}_t$ are compressible in a wavelet/Fourier basis, one can argue that the columns of C need to be compressive as well, either in a similar wavelet basis. This is also motivated by the fact that columns of C encodes the dominant motion in the scene, and for a large set of videos, this is smooth and has sparse representation in a wavelet/DCT basis or in a dictionary learnt from training data. One can exploit this along the lines of the theory of CS. However, note that $\mathbf{y}_t = C\mathbf{x}_t$ is a bilinear relationship in C and $\mathbf{x}_t$ which complicates direct inference of the unknowns. Towards alleviating this non-linearity, a two-step measurement process was proposed in [127]. It essentially allows one to estimate the state $\mathbf{x}_t$ first and subsequently solve for a sparse approximation of C. This procedure is referred to as the *CS-LDS* framework [127].

At each time instant t, the following two sets of measurements are taken:

$$\mathbf{z}_t = \begin{pmatrix} \check{\mathbf{z}}_t \\ \tilde{\mathbf{z}}_t \end{pmatrix} = \begin{bmatrix} \check{\Phi} \\ \tilde{\Phi}_t \end{bmatrix} \mathbf{y}_t = \Phi_t \mathbf{y}_t, \tag{4.25}$$

where $\check{\mathbf{z}}_t \in \mathbb{R}^{\check{M}}$ and $\tilde{\mathbf{z}}_t \in \mathbb{R}^{\tilde{M}}$, such that the total number of measurements at each frame is $M = \check{M} + \tilde{M}$. Consecutive measurements from an SPC [54] can be aggregated to provide multiple measurements at each t under the assumption of a quasi-stationary scene. Denote $\check{\mathbf{z}}_t$ as *common* measurements since the corresponding measurement matrix $\check{\Phi}$ is the same at each time instant and denote $\tilde{\mathbf{z}}$ as the *innovations* measurements.

The CS-LDS, first, solves for the state sequence $[\mathbf{x}]_{1:T}$ and subsequently, estimates the observation matrix C. The common measurements $[\check{\mathbf{z}}]_{1:T}$ are related to the state sequence $[\mathbf{x}]_{1:T}$ as follows:

$$[\check{\mathbf{z}}]_{1:T} = \begin{bmatrix} \check{\mathbf{z}}_1 & \check{\mathbf{z}}_2 & \cdots & \check{\mathbf{z}}_T \end{bmatrix} = \check{\Phi} C \begin{bmatrix} \mathbf{x}_1 & \mathbf{x}_2 & \cdots & \mathbf{x}_T \end{bmatrix} = \check{\Phi} C [\mathbf{x}]_{1:T}. \tag{4.26}$$

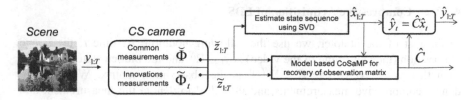

Fig. 4.8 Block diagram of the CS-LDS framework [127]

The SVD of $[\check{\mathbf{z}}]_{1:T} = USV^T$ allows us to identify $[\mathbf{x}]_{1:T}$ up to a linear transformation. In particular, the columns of V corresponding to the top d singular values form an estimate of $[\mathbf{x}]_{1:T}$ up to a $d \times d$ linear transformation (the ambiguity being the choice of coordinate). When the video sequence is exactly an LDS of d dimensions, this estimate is exact provided $\check{M} > d$. The estimate can be very accurate, when the video sequence is approximated by a d-dimensional subspace.

Once an estimate of the state sequence is obtained, say $[\widehat{\mathbf{x}}]_{1:T}$, one can obtain C by solving the following convex optimization problem:

$$(P1) \quad \min \sum_{k=1}^{d} \|\Psi^T \mathbf{c}_k\|_1, \text{ subject to } \|\mathbf{z}_t - \Phi_t C \widehat{\mathbf{x}}_t\|_2 \leq \varepsilon, \forall t \qquad (4.27)$$

where $\mathbf{c}_k$ is the k-th column of the matrix C, and Ψ is a sparsifying basis for the columns of C.

To summarize (see Fig. 4.8), the design of the measurement matrix as in (4.25) enables the estimation of the state sequence using just the common measurements, and subsequently solving for C using the diversity present in the innovations measurements $[\widetilde{\mathbf{z}}]_t$. See [127] for more details on the structure of the matrix C and a model-based greedy recovery algorithm for this problem.

4.2.3.3 Experimental Results

Figure 4.9 shows reconstruction results from data collected from a high speed camera of a candle flame. Figure 4.10 shows the estimated observation matrix as well as the state sequence. As can be seen from these figures, that the CSLDS framework is able to reconstruct the video frames as well as estimate the observation matrix well from the compressive measurements.

4.3 Shape from Gradients

Recovery of shapes and images from gradients is an important problem in many fields such as computer vision, computational photography and remote sensing. For instance, techniques such as Photometric Stereo and Shape from Shading recover

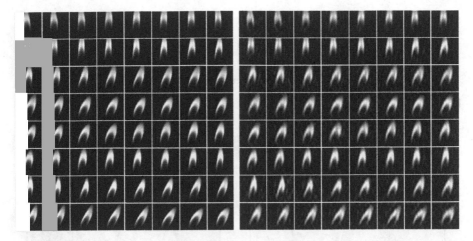

Fig. 4.9 Reconstruction of $T = 1024$ frames of a scene of resolution $N = 64 \times 64$ pixels shown as a mosaic. The original data was collected using a high speed camera operating at 1000 fps. Compressive measurements were obtained with $\check{M} = 30$ and $\tilde{M} = 20$, thereby giving a measurement rate $M/N = 1.2\%$. Reconstruction was performed using an LDS with $d - 15$ and $K = 150$. Shown above are 64 uniformly sampled frames from the ground truth (left) and the reconstruction (right) [127]

the underlying 3D shape by integrating an estimated surface gradient field or surface normals. In applications such as image stitching and image editing, gradients of given images first manipulated. The final image is then reconstructed from the modified gradient field. The estimated or modified gradient field is usually non-integrable due to the presence of noise, outliers in the estimation process and inherent ambiguities. In this chapter, we present a robust sparsity-based method for surface reconstruction from the given non-integrable gradient field [118].

4.3.1 Sparse Gradient Integration

Let $S(y,x)$ be the desired surface over a rectangular grid of size $H \times W$. Let (dx, dy) be the given non-integrable gradient field. The goal is to estimate S from (dx, dy). The integrable gradient field of S is given by the following difference equations

$$dx^0(y,x) = S(y,x+1) - S(y,x)$$
$$dy^0(y,x) = S(y+1,x) - S(y,x). \tag{4.28}$$

Let $\mathbf{s}$ be the lexicographically ordered vector representing S, $\mathbf{G}$ be the gradient operator matrix and $\mathbf{g}$ be the stacked gradinets. Each row of $\mathbf{D}$ has two non-zero entries ± 1 in pixel positions corresponding to that particular gradient. With these

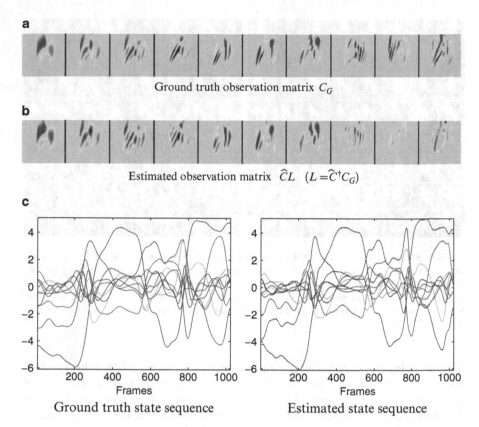

a

Ground truth observation matrix C_G

b

Estimated observation matrix $\widehat{C}L$ $(L = \widehat{C}^\dagger C_G)$

c

Ground truth state sequence   Estimated state sequence

Fig. 4.10 Ground truth and estimated parameters corresponding to Figure 4.9. Shown are the top 10 columns of the observation matrix and state sequences. Matlab's "jet" colormap (red= +large and blue= −large) is used in (a) and (b) [127]

definitions, (4.28) can be written as

$$g^0 = \begin{bmatrix} dx^0 \\ dy^0 \end{bmatrix} = \mathbf{G}s. \qquad (4.29)$$

The curl of the gradient field can be defined as the loop integral around a box of four pixels

$$\mathrm{curl}(y,x) = dx(y+1,x) - dx(y,x) + dy(y,x) - dy(y,x+1),$$

which in the matrix form can be written as follows

$$\mathbf{d} = \mathbf{C}g, \qquad (4.30)$$

where $\mathbf{d}$ denotes the vector of stacked curl values and $\mathbf{C}$ denotes the curl operator matrix.

Since the gradient field g^0 is integrable, $\mathbf{C}g^0 = 0$. However, for a general non-integrable gradient field g, $\mathbf{C}g \neq 0$. Decomposing g as the sum of g^0 and a gradient error field $\mathbf{e}$, we get

$$\mathbf{g} = \mathbf{g}^0 + \mathbf{e} = \mathbf{D}\mathbf{s} + \mathbf{e}. \tag{4.31}$$

Applying the curl operator on both sides, we obtain

$$\mathbf{d} = \mathbf{C}\mathbf{e}. \tag{4.32}$$

Hence, integrability can be defined as error correction. In other words, estimate the gradient error field $\mathbf{e}$ given the curl $\mathbf{d}$ of the corrupted gradient field. Since, there are HW known curl values and $2HW$ unknown error gradients, we have an under-determined system of linear equations.

Traditional Poisson solver finds a least squares fit to the gradients by solving the following optimization problem

$$\hat{\mathbf{e}} = \arg\min \|\mathbf{e}\|_2 \text{ subject to } \mathbf{d} = \mathbf{C}\mathbf{e}. \tag{4.33}$$

The least squares estimate is optimal when the gradient errors obey a Gaussian distribution. If the errors contains outliers, then the estimate is skewed leading to severe artifacts in the reconstructed surface or image. Outliers in the gradient field can be understood as arbitrarily large errors and could obey any distribution.

One approach to handle outliers is to combinatorially search for the possible locations of outliers, estimate them subject to the curl constraint (4.32) and pick the combination which satisfies the constraints the best. This can be done by solving the following ℓ_0-minimization problem

$$\hat{\mathbf{e}} = \arg\min \|\mathbf{e}\|_0 \text{ subject to } \mathbf{d} = \mathbf{C}\mathbf{e}. \tag{4.34}$$

However, as we saw before, the minimization of the above problem is NP-hard. Instead, one can solve the following ℓ_1-minimization problem

$$\hat{\mathbf{e}} = \arg\min \|\mathbf{e}\|_1 \text{ subject to } \mathbf{d} = \mathbf{C}\mathbf{e}. \tag{4.35}$$

Problem (4.35) is a convex problem and can be solved in polynomial time. In the presence of noise, gradient error $\mathbf{e}$ is non-sparse with the largest components corresponding to outliers. To handle noise, the cost function can be modified as follows

$$\hat{\mathbf{e}} = \arg\min \|\mathbf{e}\|_1 \text{ subject to } \|\mathbf{d} - \mathbf{C}\mathbf{e}\|_2 \le \varepsilon, \tag{4.36}$$

for an appropriate ε. Note that the recovery of $\mathbf{e}$ from the above optimization problems requires that $\mathbf{e}$ is sparse enough and $\mathbf{C}$ satisfies certain conditions [118].

4.3.2 Numerical Examples

In this section, we show some examples of reconstruction obtained by applying various methods to the problem of shape recovery from gradients. In particular, using a synthetic surface, we compare the performance of different algorithms such

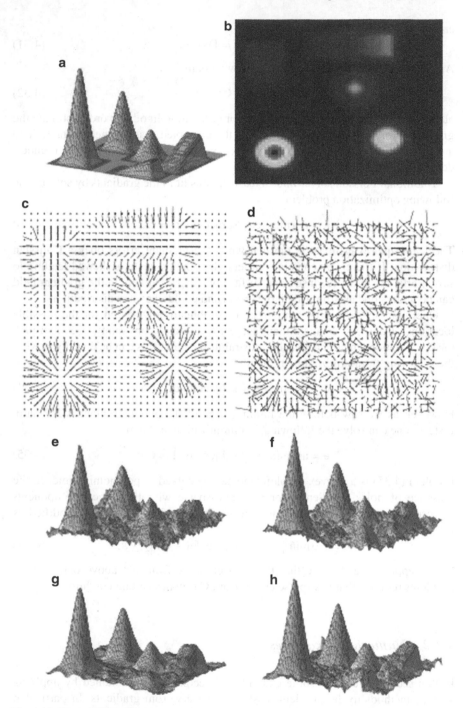

Fig. 4.11 Reconstructed surface when the gradient field is corrupted by both outliers (at 3% locations) and noise (Gaussian with $\sigma = 7\%$). (a) Original surface. (b) Image plot of the surface shown in (a). (c) Surface normal needle plot of the original surface. (d) Surface normal needle plot corresponding to the noisy gradients. (e) Reconstruction using Poisson solver. (f) Reconstruction using Frankot-Chellappa method. (g) Reconstruction using shapelets. (h) Reconstruction using ℓ_1-minimization method

as a Poisson solver-based method [134], shapelet-based approach [76], Frankot-Chellappa algorithm [62] and an ℓ_1-minimization approach [118]. We used a synthetic surface shown in Fig. 4.11(a) to generate the gradient field. We then contaminated the gradient field by adding Gaussian noise and outliers. The noisy gradient field was then used as input to different integration methods. Figure 4.11 shows the reconstructions from different methods. As can be seen from the figure, the Poisson-based method suffers the most from outliers and noise producing very poor reconstruction. The ℓ_1-minimization approach produces good results. Frankot-Chellappa method and shapelet-based approach perform significantly better compared to the Poisson-based method.

Chapter 5
Sparse Representation-based Object Recognition

In this chapter, we show how the sparse representation framework can be used to develop robust algorithms for object classification [156], [112], [106]. In particular, we will outline the Sparse Representation-based Classification (SRC) algorithm [156] and present its applications in robust biometrics recognition [156], [112], [111]. Through the use of Mercer kernels [128], we will show how SRC can be made nonlinear [163]. We will then present the algorithms for solving nonlinear sparse coding problems [41], [63], [161], [163], [130]. Finally, we will show how SRC can be generalized for multimodal multivariate sparse representations and present its application in multimodal biometrics recognition problems [130]. We first briefly outline the idea behind sparse representation[1] [17].

5.1 Sparse Representation

As we saw in Chapter 2 representing a signal involves the choice of a basis, where the signal is uniquely represented as the linear combination of the basis elements. In the case when we have the orthogonal basis, the representation coefficients are simply found by computing inner products of the signal with the basis elements. In the non-orthogonal basis case, the coefficients are found by taking the inner products of the signal with the bi-orthogonal basis. Due to the limitations of orthogonal and bi-orthogonal basis in representing complex signals, overcomplete dictionaries were developed. An overcomplete dictionary has more elements, also known as atoms, than the dimension of the signal.

Consider the dictionary $\mathbf{B} = [\mathbf{b}_1, \cdots, \mathbf{b}_L] \in \mathbb{R}^{N \times L}$, where $L \geq N$ and the columns of $\mathbf{B}$ are the dictionary atoms. Representing $\mathbf{x} \in \mathbb{R}^N$ using $\mathbf{B}$ entails solving the following optimization problem

---

[1] Also known as sparse coding.

V.M. Patel and R. Chellappa, *Sparse Representations and Compressive Sensing for Imaging and Vision*, SpringerBriefs in Electrical and Computer Engineering, DOI 10.1007/978-1-4614-6381-8_5, © The Author(s) 2013

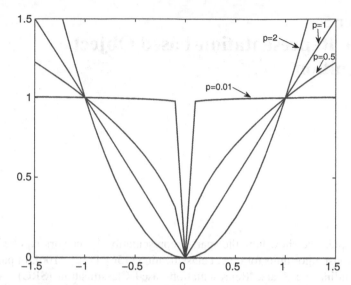

Fig. 5.1 The behavior of the scalar function $|x|^p$ for various values of p. As p goes to zeros, $|x|^p$ becomes the delta function, which is zeros for $x = 0$ and 1 elsewhere

$$\hat{\alpha} = \arg\min_{\alpha'} C(\alpha') \text{ subject to } \mathbf{x} = \mathbf{B}\alpha'. \tag{5.1}$$

for some cost function $C(\alpha)$. In practice, since we want the sorted coefficients to decay quickly, sparsity of the representation is usually enforced. This can be done by choosing, $C(\alpha) = \|\alpha\|_p$, $0 \le p \le 1$. Fig. 5.1 shows the behavior of the scalar weight functions $|\alpha|^p$ for various values of p. Note that as p goes to zero, $|\alpha|^p$ becomes a count of the nonzeros in α. Hence, by setting $C(\alpha) = \|\alpha\|_0$, one can look for the sparsest solution to the underdetermined linear system of equations $\mathbf{x} = \mathbf{B}\alpha$. The optimization problem in this case becomes the following

$$(P_0) \quad \hat{\alpha} = \arg\min_{\alpha'} \|\alpha'\|_0 \text{ subject to } \mathbf{x} = \mathbf{B}\alpha'. \tag{5.2}$$

As it turns out, this problem is NP-hard and can not be solved in polynomial time. As a result, other alternatives are usually sought. The approach often taken in practice is to instead solve the following ℓ_1-minimization problem

$$(P_1) \quad \hat{\alpha} = \arg\min_{\alpha'} \|\alpha'\|_1 \text{ subject to } \mathbf{x} = \mathbf{B}\alpha'. \tag{5.3}$$

(P_1) is a convex optimization problem and it is the one closest to (P_0) which can be solved by the standard optimization tools. Problem (5.3), is often referred to as Basis Pursuit [38]. It has been shown that for $\mathbf{B}$ with incoherent columns, whenever (P_0) has a sufficiently sparse solution, that solution is unique and is equal to the solution of (P_1).

Define the mutual coherence of the matrix $\mathbf{B}$ as follows

Definition 5.1. [17] The mutual coherence of a given matrix $\mathbf{B}$ is the largest absolute normalized inner product between different columns from $\mathbf{B}$. Denoting the kth column in $\mathbf{B}$ by $\mathbf{b}_k$, the mutual coherence is given by

$$\tilde{\mu}(\mathbf{B}) = \max_{1 \leq k,j \leq L, k \neq j} \frac{|\mathbf{b}_k^T \mathbf{b}_j|}{\|\mathbf{b}_k\|_2 \cdot \|\mathbf{b}_j\|_2}.$$

With this definition, one can prove the following theorem

Theorem 5.1. *[50], [69] For the system of linear equations* $\mathbf{x} = \mathbf{B}\alpha$ *($\mathbf{B} \in \mathbb{R}^{N \times L}$ full rank with $L \geq N$), if a solution α exists obeying*

$$\|\alpha\|_0 < \frac{1}{2}\left(1 + \frac{1}{\tilde{\mu}(\mathbf{B})}\right),$$

that solution is both unique solution of (P_1) and the unique solution of (P_0).

In the rest of the chapter we show how the variants of (5.1) can be used to develop robust algorithms for object classification.

5.2 Sparse Representation-based Classification

In object recognition, given a set of labeled training samples, the task is to identify the class to which a test sample belongs to. Following [156] and [112], in this section, we briefly describe the use of sparse representations for biometric recognition, however, this framework can be applied to a general object recognition problem.

Suppose that we are given L distinct classes and a set of n training images per class. One can extract an N-dimensional vector of features from each of these images. Let $\mathbf{B}_k = [\mathbf{x}_{k1}, \ldots, \mathbf{x}_{kj}, \ldots, \mathbf{x}_{kn}]$ be an $N \times n$ matrix of features from the kth class, where $\mathbf{x}_{kj}$ denote the feature from the jth training image of the kth class. Define a new matrix or dictionary $\mathbf{B}$, as the concatenation of training samples from all the classes as

$$\mathbf{B} = [\mathbf{B}_1, \ldots, \mathbf{B}_L] \in \mathbb{R}^{N \times (n.L)}$$
$$= [\mathbf{x}_{11}, \ldots, \mathbf{x}_{1n} | \mathbf{x}_{21}, \ldots, \mathbf{x}_{2n} | \ldots \ldots | \mathbf{x}_{L1}, \ldots, \mathbf{x}_{Ln}].$$

We consider an observation vector $\mathbf{y} \in \mathbb{R}^N$ of unknown class as a linear combination of the training vectors as

$$\mathbf{y} = \sum_{i=1}^{L} \sum_{j=1}^{n} \alpha_{ij} \mathbf{x}_{ij} \tag{5.4}$$

with coefficients $\alpha_{ij} \in \mathbb{R}$. The above equation can be written more compactly as

$$\mathbf{y} = \mathbf{B}\alpha, \tag{5.5}$$

where

$$\alpha = [\alpha_{11}, ..., \alpha_{1n} | \alpha_{21}, ..., \alpha_{2n} | ...... | \alpha_{L1}, ..., \alpha_{Ln}]^T \tag{5.6}$$

and $.^T$ denotes the transposition operation. We assume that given sufficient training samples of the kth class, $\mathbf{B}_k$, any new test image $\mathbf{y} \in \mathbb{R}^N$ that belongs to the same class will lie approximately in the linear span of the training samples from the class k. This implies that most of the coefficients not associated with class k in (5.6) will be close to zero. Hence, α is be a sparse vector.

In order to represent an observed vector $\mathbf{y} \in \mathbb{R}^N$ as a sparse vector α, one needs to solve the system of linear equations (5.5). Typically $L.n \gg N$ and hence the system of linear equations (5.5) is under-determined and has no unique solution. As we saw earlier, if α is sparse enough and $\mathbf{B}$ satisfies certain properties, then the sparsest α can be recovered by solving the following optimization problem

$$\hat{\alpha} = \arg\min_{\alpha'} \| \alpha' \|_1 \text{ subject to } \mathbf{y} = \mathbf{B}\alpha'. \tag{5.7}$$

When noisy observations are given, Basis Pursuit DeNoising (BPDN) can be used to approximate α

$$\hat{\alpha} = \arg\min_{\alpha'} \| \alpha' \|_1 \text{ subject to } \|\mathbf{y} - \mathbf{B}\alpha'\|_2 \le \varepsilon, \tag{5.8}$$

where we have assumed that the observations are of the following form

$$\mathbf{y} = \mathbf{B}\alpha + \eta \tag{5.9}$$

with $\| \eta \|_2 \le \varepsilon$.

Given an observation vector $\mathbf{y}$ from one of the L classes in the training set, one can compute its coefficients $\hat{\alpha}$ by solving either (5.7) or (5.8). One can perform classification based on the fact that high values of the coefficients $\hat{\alpha}$ will be associated with the columns of $\mathbf{B}$ from a single class. This can be done comparing how well the different parts of the estimated coefficients, $\hat{\alpha}$, represent $\mathbf{y}$. The minimum of the representation error or the residual error can then be used to identify the correct class. The residual error of class k is calculated by keeping the coefficients associated with that class and setting the coefficients not associated with class k to zero. This can be done by introducing a characteristic function, $\Pi_k : \mathbb{R}^n \to \mathbb{R}^n$, that selects the coefficients associated with the kth class as follows

$$r_k(\mathbf{y}) = \|\mathbf{y} - \mathbf{B}\Pi_k(\hat{\alpha})\|_2. \tag{5.10}$$

Algorithm 2: Sparse Representation-based Classification (SRC) Algorithm
Input: $\mathbf{D} \in \mathbb{R}^{N \times (n.L)}$, $\mathbf{y} \in \mathbb{R}^{N}$.   1. Solve the BP (5.7) or BPDN (5.8) problem.   2. Compute the residual using (5.10).   3. Identify $\mathbf{y}$ using (5.11).   **Output:** Class label of $\mathbf{y}$.

Here the vector Π_k has value one at locations corresponding to the class k and zero for other entries. The class, d, which is associated with an observed vector, is then declared as the one that produces the smallest approximation error

$$d = \arg\min_k r_k(\mathbf{y}). \tag{5.11}$$

The sparse representation-based classification method is summarized in Algorithm 2.

For classification, it is important to be able to detect and then reject the test samples of poor quality. To decide whether a given test sample has good quality, one can use the notion of Sparsity Concentration Index (SCI) proposed in [156]. The SCI of a coefficient vector $\alpha \in \mathbb{R}^{(L.n)}$ is defined as

$$SCI(\alpha) = \frac{\frac{L.\max \|\Pi_i(\alpha)\|_1}{\|\alpha\|_1} - 1}{L - 1}. \tag{5.12}$$

SCI takes values between 0 and 1. SCI values close to 1 correspond to the case where the test image can be approximately represented by using only images from a single class. The test vector has enough discriminating features of its class, so has high quality. If SCI = 0 then the coefficients are spread evenly across all classes. So the test vector is not similar to any of the classes and has poor quality. A threshold can be chosen to reject the images with poor quality. For instance, a test image can be rejected if $SCI(\hat{\alpha}) < \lambda$ and otherwise accepted as valid, where λ is some chosen threshold between 0 and 1.

5.2.1 Robust Biometrics Recognition using Sparse Representation

To illustrate the effectiveness of the SRC algorithm for face and iris biometrics, we highlight some of the results presented in [156] and [112]. The recognition rates achieved by the SRC method for face recognition with different features and dimensions are summarized in Table 5.1 on the extended Yale B Dataset [64]. As it can be seen from Table 5.1 the SRC method achieves the best recognition rate of 98.09% with randomfaces of dimension 504.

Table 5.1 Recognition Rates (in %) of SRC Algorithm [156] on the Extended Yale B Database

Dimension	30	56	120	504
Eigen	86.5	91.63	93.95	96.77
Laplacian	87.49	91.72	93.95	96.52
Random	82.60	91.47	95.53	98.09
Downsample	74.57	86.16	92.13	97.10
Fisher	86.91	-	-	-

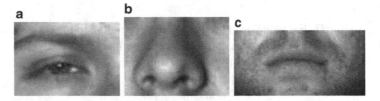

Fig. 5.2 Examples of partial facial features. (a) Eye (b) Nose (c) Mouth

Table 5.2 Recognition results with partial facial features [156]

	Right Eye	Nose	Mouth
Dimension	5,040	4,270	12,936
SRC	93.7%	87.3%	98.3%
NN	68.8%	49.2%	72.7%
NS	78.6%	83.7%	94.4%
SVM	85.8%	70.8%	95.3%

Partial face features have been very popular in recovering the identity of human face [135], [156]. The recognition results on partial facial features such as an eye, nose, and mouth are summarized in Table 5.2 on the same dataset. Examples of partial facial features are shown in Fig. 5.2. The SRC algorithm achieves the best recognition performance of $93.7\%, 87.3\%, 98.3\%$ on eye, nose and mouth features, respectively and it outperforms the other competitive methods such as Nearest Neighbor (NN), Nearest Subspace (NS) and Support Vector Machines (SVM). These results show that SRC can provide good recognition performance even in the case when partial face features are provided.

One of the main difficulties in iris biometric is that iris images acquired from a partially cooperating subject often suffer from blur, occlusion due to eyelids, and specular reflections. As a result, the performance of existing iris recognition systems degrade significantly on these images. Hence, it is essential to select good images before they are input to the recognition algorithm. To this end, one such algorithm based on SR for iris biometric was proposed in [112] that can select and recognize iris images in a single step. The block diagram of the method based on SR for iris recognition is shown in Fig. 5.3.

In Fig. 5.4, we display the iris images having the least SCI value for the blur, occlusion and segmentation error experiments performed on the real iris images in the University of Notre Dame ND dataset [14]. As it can be observed, the low SCI

Fig. 5.3 Block diagram of
the method proposed in [112]
for the selection and
recognition of iris images

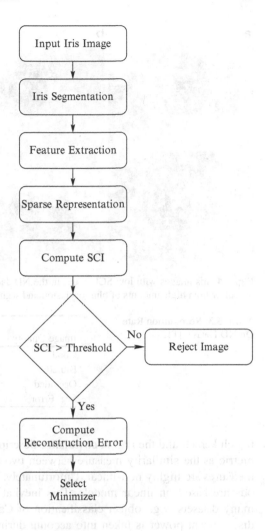

images suffer from high amounts of distortion. The recognition performance of the
SR based method for iris biometric [112] is summarized in Table 5.3. As it can be
seen from the table SRC provides the best recognition performance over that of NN
and Libor Masek's iris identification source code [89].

5.3 Non-linear Kernel Sparse Representation

Linear representations are almost always inadequate for representing nonlinear data
arising in many practical applications. For example, many types of descriptors in
computer vision have intrinsic nonlinear similarity measure functions. The most
popular ones include the spatial pyramid descriptor [78] which uses a pyramid

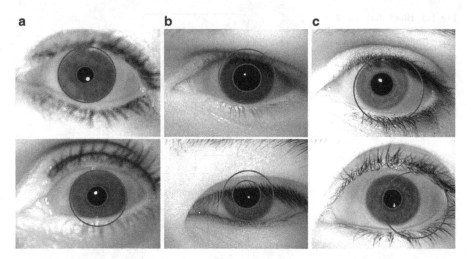

Fig. 5.4 Iris images with low SCI values in the ND dataset. Note that the images in (a), (b) and (c) suffer from high amounts of blur, occlusion and segmentation errors, respectively

Table 5.3 Recognition Rate On ND Dataset [112]

Image Quality	NN	Masek's Implementation	SRC
Good	98.33	97.5	99.17
Blured	95.42	96.01	96.28
Occluded	85.03	89.54	90.30
Seg. Error	78.57	82.09	91.36

match kernel, and the region covariance descriptor [145] which uses a Riemannian metric as the similarity measure between two descriptors. Both of these distance measures are highly non-linear. Unfortunately, the traditional sparse coding methods are based on linear models. This inevitably leads to poor performances for many datasets, e.g., object classification of Caltech-101 [73] dataset, even when discriminant power is taken into account during the training. This has motivated researchers to study non-linear kernel sparse representations [41], [63], [161], [163], [130]. In this section, we present an overview of non-linear extensions of sparse representation framework and present algorithms to solve them.

5.3.1 Kernel Sparse Coding

To make the data in an input space separable, the data is implicitly mapped into a high-dimensional kernel feature space by using some nonlinear mapping associated with a kernel function. The kernel function, $\kappa : \mathbb{R}^N \times \mathbb{R}^N \to \mathbb{R}$, is defined as the inner product

$$\kappa(\mathbf{x}_i, \mathbf{x}_j) = \langle \phi(\mathbf{x}_i), \phi(\mathbf{x}_j) \rangle, \tag{5.13}$$

where, $\phi : \mathbb{R}^N \to \mathscr{F} \subset \mathbb{R}^{\tilde{N}}$ is an implicit mapping projecting the vector $\mathbf{x}$ into a higher dimensional space, $\mathscr{F}$. Some commonly used kernels include polynomial kernels

$$\kappa(\mathbf{x}, \mathbf{y}) = \langle (\mathbf{x}, \mathbf{y}) + c \rangle^d$$

and Gaussian kernels

$$\kappa(\mathbf{x}, \mathbf{y}) = \exp\left(-\frac{\|\mathbf{x} - \mathbf{y}\|^2}{c}\right),$$

where c and d are the parameters.

By substituting the mapped features to the formulation of sparse representation, we arrive at kernel sparse representation

$$\hat{\alpha} = \arg\min_{\alpha'} \| \alpha' \|_1 \text{ subject to } \phi(\mathbf{y}) = \phi(\mathbf{B})\alpha', \tag{5.14}$$

where with the abuse of notation we denote $\phi(\mathbf{B}) = [\phi(\mathbf{b}_1), \cdots, \phi(\mathbf{b}_L)]$. In other words, kernel sparse representation seeks the sparse representation for a mapped feature under the mapped dictionary in the high dimensional feature space. Problem (5.14) can be rewritten as

$$\hat{\alpha} = \min_{\alpha'} \|\phi(\mathbf{y}) - \phi(\mathbf{B})\alpha'\|^2 + \lambda \| \alpha' \|_1, \tag{5.15}$$

where λ is a parameter and larger λ corresponds to sparser solution. The objective function in (5.15) can be simplified as

$$\min_{\alpha'} \|\phi(\mathbf{y}) - \phi(\mathbf{B})\alpha'\|^2 + \lambda \| \alpha'$$
$$= \kappa(\mathbf{y}, \mathbf{y}) + \alpha^T \mathbb{K}_{\mathbf{BB}}\alpha - 2\alpha^T \mathbb{K}_{\mathbf{B}} + \lambda \|\alpha\|_1$$
$$= g(\alpha) + \lambda \|\alpha\|_1,$$

where

$$g(\alpha) = \kappa(\mathbf{y}, \mathbf{y}) + \alpha^T \mathbb{K}_{\mathbf{BB}}\alpha - 2\alpha^T \mathbb{K}_{\mathbf{B}},$$

$\mathbb{K}_{\mathbf{BB}} \in \mathbb{R}^{L \times L}$ is a matrix with $\mathbb{K}_{\mathbf{BB}}(i, j) = \kappa(\mathbf{b}_i, \mathbf{b}_j)$ and

$$\mathbb{K}_{\mathbf{B}} \in \mathbb{R}^{L \times 1} = [\kappa(\mathbf{b}_i, \mathbf{y}), \cdots, \kappa(\mathbf{b}_L, \mathbf{y})]^T.$$

The objective is the same as that of sparse coding except for the definition of $\mathbb{K}_{\mathbf{BB}}$ and $\mathbb{K}_{\mathbf{B}}$. Hence, the standard numerical tools for solving linear sparse representation problem can be used to solve the above non-linear sparse coding problem [63].

5.3.2   Kernel Orthogonal Matching Pursuit

Similar to the ℓ_1-minimization problem for kernel sparse coding (5.14), the following problem can also be used to obtain a sparse vector in the feature space

$$\hat{\alpha} = \arg\min_{\alpha'} \|\phi(\mathbf{y}) - \phi(\mathbf{B})\alpha'\|_2 \text{ subject to } \|\alpha'\|_0 \leq C_0, \qquad (5.16)$$

where C_0 is a preset upper bound on the sparsity level. The problem (5.16) is NP-hard, which can be approximately solved by kernelizing greedy algorithms such as OMP [109]. In what follows, we describe the Kernel Orthogonal Matching Pursuit (KOMP) algorithm [41] for solving (5.16).

In KOMP, each dot product operation in OMP is replaced by the kernel trick (5.13). Using the transformed features, the dot product between $\phi(\mathbf{y})$ and a dictionary atom $\phi(\mathbf{b}_i)$ can be computed by

$$c_i = \langle \phi(\mathbf{b}_i, \phi(\mathbf{y})) \rangle = \kappa(\mathbf{b}_i, \mathbf{y}) = (\mathbb{K}_\mathbf{B})_i.$$

The orthogonal projection coefficients of $\phi(\mathbf{y})$ onto a set of selected dictionary atoms $\{\phi(\mathbf{b}_n)\}_{n \in \Lambda}$ is given as

$$\mathbf{p}_\Lambda = \left((\mathbb{K}_{\mathbf{BB}})_{\Lambda,\Lambda} \right)^{-1} (\mathbb{K}_\mathbf{B})_\Lambda.$$

The residual vector between $\phi(\mathbf{y})$ and its approximation from $\{\phi(\mathbf{b}_n)\}_{n \in \Lambda} = \phi(\mathbf{B})_{:,\Lambda}$ can be expressed as

$$\phi(\mathbf{r}) = \phi(\mathbf{y}) - \phi(\mathbf{B})_{:,\Lambda} \left((\mathbb{K}_{\mathbf{BB}})_{\Lambda,\Lambda} \right)^{-1} (\mathbb{K}_\mathbf{B})_\Lambda.$$

The representation of the residual vector $\mathbf{r}$ in the high-dimensional feature space cannot be evaluated explicitly, however the correlcation between $\phi(\mathbf{r})$ and an atom $\phi(\mathbf{b}_i)$ can be computed as follows

$$c_i = \langle \phi(\mathbf{r}), \phi(\mathbf{b}_i) \rangle = (\mathbb{K}_\mathbf{B})_i - (\mathbb{K}_{\mathbf{BB}})_{i,\Lambda} \left((\mathbb{K}_{\mathbf{BB}})_{\Lambda,\Lambda} \right)^{-1} (\mathbb{K}_\mathbf{B})_\Lambda.$$

Using these definitions, one can derive the different steps of KOMP algorithm which are summarized in Algorithm 3.

5.3.3   Kernel Simultaneous Orthogonal Matching Pursuit

Similar to the sparse coding problem (5.2), simultaneous sparse representation has also gained a lot of traction in recent years [140], [139]. In simultaneous sparse approximation, given several input signals, the objective is to approximate all these signals at once using different linear combinations of the same elementary signals, while balancing the error in approximating the data against the total number of elementary signals that are used. Let $\mathbf{X} = [\mathbf{x}_1, \cdots, \mathbf{x}_T] \in \mathbb{R}^{N \times T}$ be the matrix

---

Algorithm 3: Kernel Orthogonal Matching Pursuit (KOMP) Algorithm

---

Input: $N \times L$ dictionary $\mathbf{B} = [\mathbf{b}_1, \cdots, \mathbf{b}_L]$, sample $\mathbf{y}$, kernel function κ, and stopping criterion.
Initialize: Compute kernel matrix $\mathbb{K}_{\mathbf{BB}} \in \mathbb{R}^{L \times L}$ whose (i, j)th entry is $\kappa(\mathbf{b}_i, \mathbf{b}_j)$, and vector $\mathbb{K}_{\mathbf{B}} \in \mathbb{R}^L$ whose entries are $\kappa(\mathbf{b}_i, \mathbf{y})$. Set index set Λ_0 to be index corresponding to the largest entry in $\mathbb{K}_{\mathbf{B}}$ and iteration counter $t = 1$.
While stopping criterion has been been met do
  1. Compute the correlation vector $\mathbf{c} = [c_1, \cdots, c_L]^T$ by

$$\mathbf{c} = \mathbb{K}_{\mathbf{B}} - (\mathbb{K}_{\mathbf{BB}})_{:,\Lambda_{t-1}} \left((\mathbb{K}_{\mathbf{BB}})_{\Lambda_{t-1},\Lambda_{t-1}} \right)^{-1} (\mathbb{K}_{\mathbf{B}})_{\Lambda_{t-1}}.$$

  2. Select the new index set as

$$\lambda_t = \arg \max_{i=1,\cdots,L} |c_i|.$$

  3. Update the index set

$$\Lambda_t = \Lambda_{t-1} \bigcup \{\lambda_t\}.$$

  4. $t \leftarrow t + 1$.
Output: Index set $\Lambda = \Lambda_{t-1}$, the sparse representation $\hat{\alpha}$ whose nonzero entries indexed by Λ are $\hat{\alpha} = \left((\mathbb{K}_{\mathbf{BB}})_{\Lambda,\Lambda} \right)^{-1} (\mathbb{K}_{\mathbf{B}})_{\Lambda}$.

---

containing T input signals of dimension N. Then, the objective is to find a row-sparse matrix $\mathbf{S}$ so that $\mathbf{X} \approx \mathbf{BS}$, where $\mathbf{B}$ is the dictionary. One can recover the row-sparse matrix by solving the following optimization problem [140], [139]

$$\hat{\mathbf{S}} = \arg \max \|\mathbf{X} - \mathbf{BS}\|_F \quad \text{subject to} \quad \|\mathbf{S}\|_{\text{row},0} \leq C_0, \tag{5.17}$$

where $\|\mathbf{S}\|_{\text{row},0}$ denotes the number of non-zero rows of $\mathbf{S}$ and $\|\mathbf{X}\|_F$ is the Frobenius norm of matrix $\mathbf{X}$ defined as $\|\mathbf{X}\|_F = \sqrt{\sum_{i,j} X_{i,j}^2}$. The problem (5.17) can be approximately solved by the Simultaneous Orthogonal Matching Pursuit (SOMP) algorithm [140].

The above joint sparsity model can be extended to the feature space and the row-sparse matrix $\mathbf{S}'$ can be recovered by solving the following optimization problem

$$\hat{\mathbf{S}}' = \arg \max \|\phi(\mathbf{X}) - \phi(\mathbf{B})\mathbf{S}'\|_F \quad \text{subject to} \quad \|\mathbf{S}'\|_{\text{row},0} \leq C_0, \tag{5.18}$$

where $\phi(\mathbf{X}) = [\phi(\mathbf{x}_1), \cdots, \phi(\mathbf{x}_T)]$. Kernelized SOMP (KSOMP) can be used to solve the above optimization problem [41]. In KSOMP, at every iteration the dictionary atom that simultaneously yields the best approximation to all the T examples is selected. Let $\mathbf{C} \in \mathbb{R}^{L \times T}$ be the correlation correlation matrix whose (i, j)th entry is the correlation between $\phi(\mathbf{b}_i)$ and $\phi(\mathbf{r}_j)$, where $\phi(\mathbf{r}_j)$ is the residual vector of $\phi(\mathbf{x}_j)$. The new atom is then selected as the one associated with the row of $\mathbf{C}$ which has the maximal ℓ_p norm for some $p \geq 1$. The different steps of the algorithm for approximately solving (5.18) is summarized in Algorithm 4.

Algorithm 4: Kernel Simultaneous Orthogonal Matching Pursuit (KSOMP)
Algorithm

Input: $N \times L$ dictionary $\mathbf{B} = [\mathbf{b}_1, \cdots, \mathbf{b}_L]$, $N \times T$ data matrix $\mathbf{X}$, kernel function κ, and stopping criterion.
Initialize: Compute kernel matrix $\mathbb{K}_{\mathbf{BB}} \in \mathbb{R}^{L \times L}$ whose (i, j)th entry is $\kappa(\mathbf{b}_i, \mathbf{b}_j)$, and matrix $\mathbb{K}_{\mathbf{BX}} \in \mathbb{R}^{L \times T}$ whose (i, j)th entry is $\kappa(\mathbf{b}_i, \mathbf{x}_j)$. Set index set $\Lambda_0 = \arg\max_{i=1,\cdots,L} \|(\mathbb{K}_{\mathbf{BX}})_{i,:}\|_p$ with some $p \geq 1$ and iteration counter $t = 1$.
While stopping criterion has been been met do
　1. Compute the correlation matrix $\mathbf{C}$ by

$$\mathbf{C} = \mathbb{K}_{\mathbf{BX}} - (\mathbb{K}_{\mathbf{BB}})_{:,\Lambda_{t-1}} \left((\mathbb{K}_{\mathbf{BB}})_{\Lambda_{t-1},\Lambda_{t-1}} \right)^{-1} (\mathbb{K}_{\mathbf{BX}})_{\Lambda_{t-1},:} \in \mathbb{R}^{L \times T}.$$

　2. Select the new index set as

$$\lambda_t = \arg\max_{i=1,\cdots,L} \|\mathbf{C}_{i,:}\|_p, \ p \geq 1.$$

　3. Update the index set
$$\Lambda_t = \Lambda_{t-1} \bigcup \{\lambda_t\}.$$

　4. $t \leftarrow t+1$.
Output: Index set $\Lambda = \Lambda_{t-1}$, the sparse representation $\hat{\mathbf{S}}'$ whose nonzero rows are indexed by Λ are $\hat{\mathbf{S}}'_{\Lambda,:} = \left((\mathbb{K}_{\mathbf{BB}})_{\Lambda,\Lambda} \right)^{-1} (\mathbb{K}_{\mathbf{BX}})_{\Lambda,:}$.

5.3.4 Experimental Results

Similar to SRC, the above non-linear sparse coding formulation can be used for classification [41]. Once the sparse vector α is recovered using KOMP, the residual between the test sample and the $m$the class reconstruction in the high-dimensional feature space can be computed as

$$r_m(\mathbf{x}) = \|\phi(\mathbf{x}) - \phi(\mathbf{B})_{:,\Omega_m} \alpha_{\Omega_m}\|$$

$$= \left(\kappa(\mathbf{x},\mathbf{x}) - 2\alpha_{\Omega_m}^T (\mathbb{K}_{\mathbf{B}})_{\Omega_m} + \alpha_{\Omega_m}^T (\mathbb{K}_{\mathbf{BB}})_{\Omega_m,\Omega_m} \alpha_{\Omega_m} \right)^{\frac{1}{2}},$$

where Ω_m is the index set associated with the mth class. The class label of $\mathbf{x}$ can then be determined by

$$class(\mathbf{x}) = \arg\min_{m=1,\cdots,L} r_m(\mathbf{x}).$$

Similarly, in the case of *KSOMP*, once the matrix $\mathbf{S}$ is recovered, the total residual between the T samples and their approximations from the mth class training samples is given by

$$r_m(\mathbf{x}_1) = \left(\sum_{i=1}^{T} \left(\kappa(\mathbf{x}_i,\mathbf{x}_i) - 2\mathbf{S}_{\Omega_m,i}^T (\mathbb{K}_{\mathbf{BX}})_{\Omega_m,i} + \mathbf{S}_{\Omega_m,i}^T (\mathbb{K}_{\mathbf{BB}})_{\Omega_m,\Omega_m} \mathbf{S}_{\Omega_m,i} \right) \right)^{\frac{1}{2}}.$$

Table 5.4 Classification accuracy (%) for the University of Pavia dataset [41]

Class	SVM	SVMCK	KLR	KLRCK	OMP	KOMP	SOMP	KSOMP
1	84.30	79.85	82.96	74.40	68.23	76.09	59.33	94.23
2	67.01	84.86	83.34	85.91	67.04	69.61	78.15	76.74
3	68.43	81.87	64.13	61.71	65.45	72.12	83.53	79.23
4	97.80	96.36	96.33	96.22	97.29	98.11	96.91	95.12
5	99.37	99.37	99.19	99.10	99.73	99.73	99.46	100
6	92.45	93.55	80.05	84.45	73.27	87.66	77.41	99.50
7	89.91	90.21	84.51	85.32	87.26	88.07	98.57	99.80
8	92.42	92.81	83.17	93.37	81.87	89.51	89.09	98.78
9	97.23	95.35	89.81	96.48	95.97	93.96	91.95	29.06
Overall	79.15	87.18	83.56	84.77	73.30	78.33	79.00	85.67
Average	87.66	90.47	84.83	86.33	81.79	86.10	86.04	85.83

The class label for the sample $\mathbf{x}_1$ is then given by

$$class(\mathbf{x}_1) = \arg \min_{m=1,\cdots,L} r_m(\mathbf{x}_1).$$

This framework for classification was successfully applied to the hyperspectral classification problem in [41]. We highlight some of the results on the University of Pavia image using the non-linear sparse coding methods. The image consists of 1096×492 pixels, each having 102 spectral bands. About 5% of the labeled data are used as training samples. The classification results are summarized in Table 5.4. As can be seen from this table, that operating in the feature space significantly improves the accuracy of sparse coding methods on this dataset.

The ideas presented in this section can be extended to the case of multi-modal multivariate sparse representation which is covered in the next section. For simplicity, we present the multivariate sparse representation framework in terms of multimodal biometrics recognition [130], however, it can be used for any multimodal or multichannel classification problem [97].

5.4 Multimodal Multivariate Sparse Representation

Consider a multimodal C-class classification problem with D different biometric traits. Suppose there are p_i training samples in each biometric trait. For each biometric trait $i = 1, \ldots, D$, we denote

$$\mathbf{X}^i = [\mathbf{X}_1^i, \mathbf{X}_2^i, \ldots, \mathbf{X}_C^i]$$

as an $n_i \times p_i$ dictionary of training samples consisting of C sub-dictionaries $\mathbf{X}_k^i$'s corresponding to C different classes. Each sub-dictionary

$$\mathbf{X}_j^i = [\mathbf{x}_{j,1}^i, \mathbf{x}_{j,2}^i, \ldots, \mathbf{x}_{j,p_j}^i] \in \mathbb{R}^{n \times p_j}$$

represents a set of training data from the ith modality labeled with the jth class. Note that n_i is the feature dimension of each sample and there are p_j number of training samples in class j. Hence, there are a total of $p = \sum_{j=1}^{C} p_j$ many samples in the dictionary $\mathbf{X}_C^i$. Elements of the dictionary are often referred to as atoms. In multimodal biometrics recognition problem given test samples (a matrix) $\mathbf{Y}$, which consists of D different modalities $\{\mathbf{Y}^1, \mathbf{Y}^2, \ldots, \mathbf{Y}^D\}$ where each sample $\mathbf{Y}^i$ consists of d_i observations $\mathbf{Y}^i = [\mathbf{y}_1^i, \mathbf{y}_2^i, \ldots, \mathbf{y}_d^i] \in \mathbb{R}^{n \times d_i}$, the objective is to identify the class to which a test sample $\mathbf{Y}$ belongs to. In what follows, we present a multimodal multivariate sparse representation-based algorithm for this problem [130], [160], [90], [97].

5.4.1   Multimodal Multivariate Sparse Representation

We want to exploit the joint sparsity of coefficients from different biometrics modalities to make a joint decision. To simplify this model, let us consider a bimodal classification problem where the test sample $\mathbf{Y} = [\mathbf{Y}^1, \mathbf{Y}^2]$ consists of two different modalities such as iris and face. Suppose that $\mathbf{Y}^1$ belongs to the jth class. Then, it can be reconstructed by a linear combination of the atoms in the sub-dictionary $\mathbf{X}_j^1$. That is, $\mathbf{Y}^1 = \mathbf{X}^1 \Gamma^1 + \mathbf{N}^1$, where Γ^1 is a sparse matrix with only p_j nonzero rows associated with the jth class and $\mathbf{N}^1$ is noise matrix. Similarly, since $\mathbf{Y}^2$ represents the same even, it belongs to the same class and can be represented by training samples in $\mathbf{X}_j^2$ with different set of coefficients Γ_j^2. Thus, we can write $\mathbf{Y}^2 = \mathbf{X}^2 \Gamma^2 + \mathbf{N}^2$, where Γ^2 is a sparse matrix that has the same sparsity pattern as Γ^1. If we let $\Gamma = [\Gamma^1, \Gamma^2]$, then Γ is a sparse matrix with only p_j nonzeros rows.

In the more general case where we have D modalities, if we denote $\{\mathbf{Y}^i\}_{i=1}^D$ as a set of D observations each consisting of d_i samples from each modality and let $\Gamma = [\Gamma^1, \Gamma^2, \ldots, \Gamma^D] \in \mathbb{R}^{p \times d}$ be the matrix formed by concatenating the coefficient matrices with $d = \sum_{i=1}^{D} d_i$, then we can seek for the row-sparse matrix Γ by solving the following ℓ_1/ℓ_q-regularized least square problem

$$\hat{\Gamma} = \arg\min_{\Gamma} \frac{1}{2} \sum_{i=1}^{D} \|\mathbf{Y}^i - \mathbf{X}^i \Gamma^i\|_F^2 + \lambda \|\Gamma\|_{1,q} \tag{5.19}$$

where λ is a positive parameter and q is set greater than 1 to make the optimization problem convex. Here, $\|\Gamma\|_{1,q}$ is a norm defined as $\|\Gamma\|_{1,q} = \sum_{k=1}^{p} \|\gamma^k\|_q$ where γ^k's are the row vectors of Γ. Once $\hat{\Gamma}$ is obtained, the class label associated to an observed vector is then declared as the one that produces the smallest approximation error

$$\hat{j} = \arg\min_{j} \sum_{i=1}^{D} \|\mathbf{Y}^i - \mathbf{X}^i \delta_j^i(\Gamma^i)\|_F^2, \tag{5.20}$$

where δ^i_j is the matrix indicator function defined by keeping rows corresponding to the jth class and setting all other rows equal to zero. Note that the optimization problem (5.19) reduces to the conventional Lasso [137] when $D = 1$ and $d = 1$. The resulting classification algorithm reduces to SRC [156]. In the case, when $D = 1$ (5.19) is referred to as multivariate Lasso [160].

5.4.2   Robust Multimodal Multivariate Sparse Representation

In this section, we consider a more general problem where the data is contaminated by noise. In this case, the observation model can be modeled as

$$\mathbf{Y}^i = \mathbf{X}^i \Gamma^i + \mathbf{Z}^i + \mathbf{N}^i, \quad i = 1, \ldots D, \tag{5.21}$$

where $\mathbf{N}^i$ is a small dense additive noise and $\mathbf{Z}^i \in \mathbb{R}^{n \times d_i}$ is a matrix of background noise (occlusion) with arbitrarily large magnitude. One can assume that each $\mathbf{Z}^i$ is sparsely represented in some basis $\mathbf{B}^i \in \mathbb{R}^{n \times m^i}$. That is, $\mathbf{Z}^i = \mathbf{B}^i \Lambda^i$ for some sparse matrices $\Lambda^i \in \mathbb{R}^{m_i \times d_i}$. Hence, (5.21) can be rewritten as

$$\mathbf{Y}^i = \mathbf{X}^i \Gamma^i + \mathbf{B}^i \Lambda^i + \mathbf{N}^i, \quad i = 1, \ldots D, \tag{5.22}$$

With this model, one can simultaneously recover the coefficients Γ^i and as well as Λ^i by taking the advantage of that fact that Λ^i are sparse

$$\hat{\Gamma}, \hat{\Lambda} = \arg\min_{\Gamma, \Lambda} \frac{1}{2} \sum_{i=1}^{D} \|\mathbf{Y}^i - \mathbf{X}^i \Gamma^i - \mathbf{B}^i \Lambda^i\|_F^2 + \lambda_1 \|\Gamma\|_{1,q} + \lambda_2 \|\Lambda\|_1,$$

where λ_1 and λ_2 are positive parameters and $\Lambda = [\Lambda^1, \Lambda^2, \ldots, \Lambda^D]$ is the sparse coefficient matrix corresponding to occlusion. The ℓ_1-norm of matrix Λ is defined as $\|\Lambda\|_1 = \sum_{i,j} |\Lambda_{i,j}|$. Note that the idea of exploiting the sparsity of occlusion term has been studied by Wright et al. [156] and Candes et al. [27].

Once Γ, Λ are computed, the effect of occlusion can be removed by setting $\tilde{\mathbf{Y}}^i = \mathbf{Y}^i - \mathbf{B}^i \Lambda^i$. One can then declare the class label associated to an observed vector as

$$\hat{j} = \arg\min_j \sum_{i=1}^{D} \|\mathbf{Y}^i - \mathbf{X}^i \delta^i_j(\Gamma^i) - \mathbf{B}^i \Lambda^i\|_F^2. \tag{5.23}$$

Alternating Direction Method of Multipliers (ADMM) can be used to solve the above optimization algorithms. Details can be found in [130]. The algorithm for multimodal biometrics recognition is summarized in Algorithm 5.

Algorithm 5: Sparse Multimodal Biometrics Recognition (SMBR)

Input: Training samples $\{\mathbf{X_i}\}_{i=1}^D$, test sample $\{\mathbf{Y_i}\}_{i=1}^D$, Occlusion basis $\{\mathbf{B}\}_{i=1}^D$
Procedure: Obtain $\hat{\Gamma}$ and $\hat{\Lambda}$ by solving

$$\hat{\Gamma}, \hat{\Lambda} = \arg\min_{\Gamma, \Lambda} \frac{1}{2} \sum_{i=1}^D \|\mathbf{Y}^i - \mathbf{X}^i \Gamma^i - \mathbf{B}^i \Lambda^i\|_F^2 + \lambda_1 \|\Gamma\|_{1,q} + \lambda_2 \|\Lambda\|_1,$$

Output: $\texttt{identity}(\mathbf{Y}) = \arg\min_j \sum_{i=1}^D \|\mathbf{Y}^i - \mathbf{X}^i \delta_j^i(\hat{\Gamma}^i) - \mathbf{B}^i \hat{\Lambda}^i\|_F^2.$

Table 5.5 WVU Biometric Data

Biometric Modality	# of subjects	# of samples
Iris	244	3099
Fingerprint	272	7219
Palm	263	683
Hand	217	3062
Voice	274	714

Fig. 5.5 Examples of challenging images from the WVU Multimodal dataset. The images above suffer from various artifacts as sensor noise, blur, occlusion and poor acquisition.

5.4.3 Experimental Results

In this section, we highlight some of the results [130] of SMBR on the WVU dataset [124]. The WVU dataset is one of the few publicly available datasets which allows fusion at image level. It is a challenging dataset having samples from differnent biometric modalities for each subject. The WVU multimodal dataset is a comprehensive collection of different biometric modalities such as fingerprint, iris, palmprint, hand geometry and voice from subjects of different age, gender and ethnicity as described in Table 5.5. It is a challenging dataset and many of these samples are of poor quality corrupted with blur, occlusion and sensor noise as shown in Fig. 5.5. Out of these, we chose iris and fingerprint modalities for testing the algorithm. Also, the evaluation was done on the subset of subjects having samples in both the modalities. In all the experiments $\mathbf{B}_i$ was set to be identity for convenience.

5.4.3.1   Preprocessing

Robust pre-processing of images was done before feature extraction. Iris images were segmented following the recent method proposed in [113]. Following the segmentation, 25×240 iris templates were formed by re-sampling using the publicly available code of Masek *et al.* [89]. Fingerprint images were enhanced using the filtering based methods described in [123], and then the core point was detected using the enhanced images [72]. Features were then extracted around the detected core point.

5.4.3.2   Feature Extraction

Gabor features were extracted on the processed images as they have been shown to give good performance on both fingerprints [72] and iris [46]. For fingerprint samples, the processed images were convolved with Gabor filters at 8 different orientations. Circular tesselations were extracted around the core point for all the filtered images similar to [72]. The tesselation consisted of 15 concentric bands, each of width 5 pixels and divided into 30 sectors. The mean values for each sector were concatenated to form the feature vector of size 3600×1. Features for iris images were formed by convolving the templates with log-Gabor filter at a single scale, and vectorizing the template to give 6000×1 dimensional feature.

5.4.3.3   Experimental Set-up

The extracted features were randomly divided into 4 training samples per class and rest testing for each modality. The recognition result was averaged over a few runs. The proposed methods were compared with state-of-the-art classification methods such as sparse logistic regression (SLR) [77] and SVM [18]. Although these methods have been shown to give superior performance, they cannot handle multimodal data. One possible way to handle it is using feature concatenation. But, this resulted in feature vectors of size 26400×1 when all modalities are used, and hence, is not useful. Hence, two techniques were explored for fusion of results for individual modality. In the first technique, a score-based fusion was followed where the probability outputs for test sample of each modality, $\{\mathbf{y}_i\}_{i=1}^6$ were added together to give a final score vector. Classification was based upon the final score values. For the second technique, the subject chosen by the maximum number of modalities was taken to be from the correct class.

The recognition performances of SMBR-WE and SMBR-E was compared with linear SVM and linear SLR classification methods. The parameter values λ_1 and λ_2 were set to 0.01 experimentally. Table 5.6 demonstrate the recognition performance of different methods. Clearly, the SMBR approach outperforms existing classification techniques. Both SMBR-E and SMBR-WE have similar performance, though the latter seems to give a slightly better performance. This may be due to the penalty

Table 5.6 Rank one recognition performance for WVU Multimodal dataset [130]

	SMBR-WE	SMBR-E	SLR-Sum	SLR-Major	SVM-Sum	SVM-Major
4 Fingerprints	**97.9**	97.6	96.3	74.2	90.0	73.0
2 Irises	76.5	**78.2**	72.7	64.2	62.8	49.3
Overall	**98.7**	98.6	97.6	84.4	94.9	81.3

on the sparse error, though the error may not be sparse in image domain. Further, sum-based fusion shows a superior performance over voting-based methods. A surprising result is the performance of SLR using all the modalities, which is lower than its performance on the fingerprints. Due to the poor quality of iris images, the performance of sum and voting based fusion techniques go down. However, by jointly classifying all the modalities together, SMBR achieves a robust performance, even though no weights based on quality has been assigned during testing.

5.5   Kernel Space Multimodal Recognition

The class identities in the multibiometric dataset may not be linearly separable. Hence, one can also extend the sparse multimodal fusion framework to kernel space.

5.5.1   Multivariate Kernel Sparse Representation

Considering the general case of D modalities with $\{\mathbf{Y}^i\}_{i=1}^{D}$ as a set of d_i observations, the feature space representation can be written as:

$$\Phi(\mathbf{Y^i}) = [\phi(\mathbf{y}_1^i), \phi(\mathbf{y}_2^i), ..., \phi(\mathbf{y}_d^i)]$$

Similarly, the dictionary of training samples for modality $i = 1, \cdots, D$ can be represented in feature space as

$$\Phi(\mathbf{X}^i) = [\phi(\mathbf{X}_1^i), \phi(\mathbf{X}_2^i), \cdots, \phi(\mathbf{X}_C^i)]$$

As in the joint linear space representation, we have:

$$\Phi(\mathbf{Y}^i) = \Phi(\mathbf{X}^i)\Gamma^i$$

where, Γ^i is the coefficient matrix associated with modality i. Incorporating information from all the sensors, we seek to solve the following optimization problem similar to the linear case:

$$\hat{\Gamma} = \arg\min_{\Gamma} \frac{1}{2} \sum_{i=1}^{D} \|\Phi(\mathbf{Y}^i) - \Phi(\mathbf{X}^i)\Gamma^i\|_F^2 + \lambda \|\Gamma\|_{1,q} \tag{5.24}$$

where, $\Gamma = [\Gamma^1, \Gamma^2, \cdots, \Gamma^D]$. It is clear that the information from all modalities are integrated via the shared sparsity sparsity pattern of the matrices $\{\Gamma^i\}_{i=1}^D$. This can be reformulated in terms of kernel matrices as:

$$\hat{\Gamma} = \arg\min_{\Gamma} \frac{1}{2} \sum_{i=1}^{D} \left(\text{trace}(\Gamma^{i^T} \mathbf{K}_{\mathbf{X}_i, \mathbf{X}_i} \Gamma^i) - 2\text{trace}(\mathbf{K}_{\mathbf{X}_i, \mathbf{Y}_i} \Gamma^i) \right) + \lambda \|\Gamma\|_{1,q}$$

where, the kernel matrix $\mathbf{K}_{\mathbf{A},\mathbf{B}}(i,j) = \langle \phi(\mathbf{a}_i), \phi(\mathbf{b}_j) \rangle$, $\mathbf{a}_i$ and $\mathbf{b}_j$ being i^{th} and j^{th} columns of $\mathbf{A}$ and $\mathbf{B}$ respectively.

5.5.2 Composite Kernel Sparse Representation

Another way to combine information of different modalities is through composite kernel, which efficiently combines kernel for each modality. The kernel combines both within and between similarities of different modalities. For two modalities with same feature dimension, the kernel matrix can be constructed as:

$$\kappa(\mathbf{X}_i, \mathbf{X}_j) = \alpha_1 \kappa(\mathbf{x}_i^1, \mathbf{x}_j^1) + \alpha_2 \kappa(\mathbf{x}_i^1, \mathbf{x}_j^2) + \alpha_3 \kappa(\mathbf{x}_i^2, \mathbf{x}_j^1) + \alpha_4 \kappa(\mathbf{x}_i^2, \mathbf{x}_j^2)$$

where, $\{\alpha_i\}_{i=1,\cdots,4}$ are weights for kernels and $\mathbf{X}_i = [\mathbf{x}_i^1; \mathbf{x}_i^2]$, $\mathbf{x}_i^1$ being the feature vector. It can be similarly extended to multiple modalities. However, the modalities may be of different dimensions. In such cases, cross-simlarity measure is not possible. Hence, the modalities are divided according to being homogenous (e.g. right and left iris) or heterogenous (fingerprint and iris). This is also reasonable, because homogenous modalities are correlated at feature level but heteregenous modalities may not be correlated. For D modalities, with $\{d_i\}_{i \in \mathscr{S}_j}$, $\mathscr{S}_j \subseteq \{1, 2, \cdots, D\}$ being the sets of indices of each homogenous modality, the composite kernel for each set is given as:

$$\kappa(\mathbf{X}_i^k, \mathbf{X}_j^k) = \sum_{s_1 s_2 \in \mathscr{S}_k} \alpha_{s_1 s_2} \kappa(\mathbf{x}_i^{s_1}, \mathbf{x}_j^{s_2}) \tag{5.25}$$

Here, $\mathbf{X}_i^k = [\mathbf{x}_i^{s_1}; \mathbf{x}_i^{s_2}; \cdots; \mathbf{x}_i^{s_{|\mathscr{S}_k|}}]$, $\mathscr{S}_k = [s_1, s_2, \cdots, s_{|\mathscr{S}_k|}]$ and $k = 1, \cdots, N_H$, N_H being the number of different heterogenous modalities. The information from the different heterogenous modalities can then be combined similar to sparse kernel fusion case:

$$\hat{\Gamma} = \arg\min_{\Gamma} \frac{1}{2} \sum_{i=1}^{N_H} \left(\text{trace}(\Gamma^{i^T} \mathbf{K}_{\mathbf{X}^i, \mathbf{X}^i} \Gamma^i) - 2\text{trace}(\mathbf{K}_{\mathbf{X}^i, \mathbf{Y}^i} \Gamma^i) \right) + \lambda \|\Gamma\|_{1,q}$$

where, $\mathbf{K}_{\mathbf{X}^i, \mathbf{X}^i}$ is defined for each $\mathscr{S}_i$ as in previous section and $\Gamma = [\Gamma^1, \Gamma^2, \cdots, \Gamma^{N_H}]$.

Algorithm 6: Kernel Sparse Multimodal Biometrics Recognition (kerSMBR)

Input: Training samples $\{\mathbf{X_i}\}_{i=1}^{D}$, test sample $\{\mathbf{Y_i}\}_{i=1}^{D}$
Procedure: Obtain $\hat{\Gamma}$ by solving

$$\hat{\Gamma} = \arg\min_{\Gamma} \frac{1}{2} \sum_{i=1}^{D} \|\Phi(\mathbf{Y}^i) - \Phi(\mathbf{X}^i)\Gamma^i\|_F^2 + \lambda\|\Gamma\|_{1,q} \qquad (5.26)$$

Output:
$\texttt{identity}(\mathbf{Y}) = \arg\min_j \Sigma_{i=1}^{D} \left(\text{trace}(\mathbf{K_{YY}}) - 2\text{trace}(\hat{\Gamma}_j^{i^T} \mathbf{K_{X^iY}}\hat{\Gamma}_j^i) + \text{trace}(\hat{\Gamma}_j^{i^T} \mathbf{K_{X_j^iX_j^i}}\hat{\Gamma}_j^i) \right)$

Algorithm 7: Composite Kernel Sparse Multimodal Biometrics Recognition (compSMBR)

Input: Training samples $\{\mathbf{X_i}\}_{i=1}^{D}$, test sample $\{\mathbf{Y_i}\}_{i=1}^{D}$
Procedure: Obtain $\hat{\Gamma}$ by solving

$$\hat{\Gamma} = \arg\min_{\Gamma} \frac{1}{2} \sum_{i=1}^{N_H} \left(\text{trace}(\Gamma^{i^T} \mathbf{K_{X^i,X^i}}\Gamma^i) - 2\text{trace}(\mathbf{K_{X^i,Y^i}}\Gamma^i) \right) + \lambda\|\Gamma\|_{1,q}$$

Output:
$\texttt{identity}(\mathbf{Y}) = \arg\min_j \Sigma_{i=1}^{N_H} \left(\text{trace}(\mathbf{K_{YY}}) - 2\text{trace}(\hat{\Gamma}_j^{i^T} \mathbf{K_{X_j^iY}}\hat{\Gamma}_j^i) + \text{trace}(\hat{\Gamma}_j^{i^T} \mathbf{K_{X_j^iX_j^i}}\hat{\Gamma}_j^i) \right)$

Similar to the linear fusion method, the alternating direction method can be used to efficiently solve the problem for kernel fusion [130]. Multivariate kernel sparse recognition and composite kernel sparse recognition algorithms are summarized in Algorithm 6 and Algorithm 7, respectively.

5.5.3   Experimental Results

The performances of kerSMBR and compSMBR with kernel SVM and kernel SLR methods were evaluated on the WVU dataset. The Radial Basis Function (RBF) was used as the kernel, which is given as:

$$\kappa(\mathbf{x}_i, \mathbf{x}_j) = \exp\left(-\frac{\|\mathbf{x}_i - \mathbf{x}_j\|_2^2}{\sigma^2} \right),$$

σ being a parameter to control the width of the RBF. The weights $\{\alpha_{ij}\}$ were set to 1 for composite kernel. λ and β_W were set to 0.01 and 0.01 respectively.

Table 5.7 shows the performance of different methods on different fusion settings [130]. kerSMBR achieves the best accuracy among all the methods. kerSLR scores over kerSVM in all the cases, and its accuracy is close to kerSMBR.

Table 5.7 Rank one recognition performance for WVU Multimodal dataset [130].

	kerSMBR	kerSLR-Sum	kerSLR-Major	kerSVM-Sum	kerSVM-Major	compSMBR	compSLR-Sum	compSVM-Sum
4 Fingerprints	**97.9**	96.8	75.3	93.2	71.4	93.4	95.7	81.7
2 Irises	**84.7**	83.8	75.2	62.2	47.8	78.9	78.9	55.8
Overall	**99.1**	98.9	87.9	96.3	79.5	95.9	98.2	90.4

The performance of kerSMBR and kerSLR are better than their linear counterparts, however kerSVM does not show much improvement. *Composite kernels* present an interesting case. Here, compSLR shows better performance than compSMBR on all the modalities. Composite kernel by combining homogenous modalities into one, reduces effective number of modalities, hence the size of Γ matrix is reduced. This decreases the flexiblity in exploiting different modality information via Γ. Hence, the performance of compSMBR is not optimal.

Chapter 6
Dictionary Learning

Instead of using a pre-determined dictionary $\mathbf{B}$, as in (5.1), one can directly learn it from the data [99]. Indeed, it has been observed that learning a dictionary directly from the training data rather than using a predetermined dictionary usually leads to better representation and hence can provide improved results in many practical image processing applications such as restoration and classification [121], [155], [100], ,[40], [107], [131], [133]. In this section, we will highlight some of the methods for learning dictionaries and present their applications in object representation and classification.

6.1 Dictionary Learning Algorithms

Several algorithms have been developed for the task of learning a dictionary. Two of the most well-known algorithms are the Method of Optimal Directions (MOD) [58] and the KSVD algorithm [2]. Given a set of examples $\mathbf{X} = [\mathbf{x}_1, \cdots, \mathbf{x}_n]$, the goal of the KSVD and MOD algorithms is to find a dictionary $\mathbf{B}$ and a sparse matrix Γ that minimize the following representation error

$$(\hat{\mathbf{B}}, \hat{\Gamma}) = \arg\min_{\mathbf{B}, \Gamma} \|\mathbf{X} - \mathbf{B}\Gamma\|_F^2 \text{ subject to } \|\gamma_i\|_0 \leq T_0 \ \forall i,$$

where γ_i represent the columns of Γ and T_0 denotes the sparsity level. Both MOD and KSVD are iterative methods and alternate between sparse-coding and dictionary update steps. First, a dictionary $\mathbf{B}$ with ℓ_2 normalized columns is initialized. Then, the main iteration is composed of the following two stages:

- *Sparse coding*: In this step, $\mathbf{B}$ is fixed and the following optimization problem is solved to compute the representation vector γ_i for each example $\mathbf{x}_i$

$$i = 1, \cdots, n, \quad \min_{\gamma_i} \|\mathbf{x}_i - \mathbf{B}\gamma_i\|_2^2 \text{ s. t. } \|\gamma_i\|_0 \leq T_0.$$

V.M. Patel and R. Chellappa, *Sparse Representations and Compressive Sensing for Imaging and Vision*, SpringerBriefs in Electrical and Computer Engineering, DOI 10.1007/978-1-4614-6381-8_6, © The Author(s) 2013

Approximate solutions of the above problem can obtained by using any sparse coding algorithms such OMP [109], [138] and BP [38].

- *Dictionary update*: This is where the MOD algorithm differs from the KSVD algorithm. The MOD algorithm updates all the atoms simultaneously by solving the following optimization problem

$$\arg\min_{\mathbf{B}} \|\mathbf{X} - \mathbf{B}\Gamma\|_F^2 \qquad (6.1)$$

whose solution is given by

$$\mathbf{B} = \mathbf{X}\Gamma^T \left(\Gamma\Gamma^T\right)^{-1}.$$

Even though the MOD algorithm is very effective and usually converges in a few iterations, it suffers from the high complexity of the matrix inversion step as discussed in [2].

In the case of KSVD, the dictionary update is performed atom-by-atom in an efficient way rather than using a matrix inversion. The term to be minimized in equation (6.1) can be rewritten as

$$\|\mathbf{X} - \mathbf{B}\Gamma\|_F^2 = \left\|\mathbf{X} - \sum_j \mathbf{b}_j \gamma_j^T\right\|_2^2 = \left\|\left(\mathbf{X} - \sum_{j \neq j_0} \mathbf{b}_j \gamma_j^T\right) - \mathbf{b}_{j_0} \gamma_{j_0}^T\right\|_2^2,$$

where γ_j^T represents the jth row of Γ. Since we want to update $\mathbf{b}_{j_0}$ and $\gamma_{j_0}^T$, the first term in the above equation

$$\mathbf{E}_{j_0} = \mathbf{X} - \sum_{j \neq j_0} \mathbf{b}_j \gamma_j^T$$

can be precomputed. The optimal $\mathbf{b}_{j_0}$ and $\gamma_{j_0}^T$ are found by an SVD decomposition. In particular, while fixing the cardinalities of all representations, a subset of the columns of $\mathbf{E}_{j_0}$ are taken into consideration. This way of updating leads to a substantial speedup in the convergence of the training algorithm compared to the MOD method.

Both the MOD and the KSVD dictionary learning algorithms are described in detail in Algorithm 8.

6.2 Discriminative Dictionary Learning

Dictionaries can be trained for both reconstruction and discrimination applications. In the late nineties, Etemand and Chellappa proposed a linear discriminant analysis (LDA) based basis selection and feature extraction algorithm for classification using

---

Algorithm 8: The MOD and KSVD dictionary learning algorithms

Objective: Find the best dictionary to represent the samples $\mathbf{X} = [\mathbf{x}_1, \cdots, \mathbf{x}_n]$ as sparse compositions, by solving the following optimization problem:

$$\arg\min_{\mathbf{B}, \Gamma} \|\mathbf{X} - \mathbf{B}\Gamma\|_F^2 \; subject \; to \; \forall i \; \|\gamma_i\|_0 \leq T_0.$$

Input: Initial dictionary $\mathbf{B}^{(0)} \in \mathbb{R}^{N \times P}$, with normalized columns, signal matrix
$\mathbf{X} = [\mathbf{x}_1, \cdots, \mathbf{x}_n]$ and sparsity level T_0.
  1. *Sparse coding stage*:
Use any pursuit algorithm to approximate the solution of

$$\hat{\gamma}_i = \arg\min \|\mathbf{x}_i - \mathbf{B}\gamma\|_2^2 \; subject \; to \; \|\gamma\|_0 \leq T_0$$

obtaining sparse representation vector $\hat{\gamma}_i$ for $1 \leq i \leq n$. These form the matrix Γ.
  2. *Dictionary update stage*:
MOD: Update the dictionary by the formula

$$\mathbf{B} = \mathbf{X}\Gamma^T \left(\Gamma\Gamma^T\right)^{-1}.$$

KSVD: For each column $k = 1, \cdots, P$ in $\mathbf{B}^{(J-1)}$ update by
- Define the group of examples that use this atom,

$$\omega_k = \{i | 1 \leq i \leq P, \gamma_k^T(i) \neq 0\}.$$

- Compute the overall representation error matrix, $\mathbf{E}_k$, by

$$\mathbf{E}_k = \mathbf{X} - \sum_{j \neq k} \mathbf{b}_j \gamma_j^T.$$

- Restrict $\mathbf{E}_k$ by choosing only the columns corresponding to ω_k and obtain $\mathbf{E}_k^R$.
- Apply SVD decomposition $\mathbf{E}_k^R = \mathbf{U}\Delta\mathbf{V}^T$. Select the updated dictionary column
$\hat{\mathbf{b}}_k$ to be the first column of $\mathbf{U}$. Update the coefficient vector γ_R^k to be the first column
of $\mathbf{V}$ multiplied by $\Delta(1,1)$.
  3. Set $J = J + 1$.
Output: Trained dictionary $\mathbf{B}$ and sparse coefficient matrix Γ.

---

wavelet packets [59]. Recently, similar algorithms for simultaneous sparse signal representation and discrimination have also been proposed in [71, 75, 81, 101, 119]. The basic idea in learning a discriminative dictionary is to add a Linear Discriminant Analysis (LDA) type of discrimination on the sparse coefficients which essentially enforces separability among dictionary atoms of different classes. Some of the other methods for learning discriminative dictionaries include [73, 81–84, 164]. Additional techniques may be found within these references.

In particular, a dictionary learning method based on information maximization principle was proposed in [115] for action recognition. The objective function in [115] maximizes the mutual information between what has been learned and what remains to be learned in terms of appearance information and class distribution for each dictionary item. A Gaussian Process (GP) model is proposed for sparse representation to optimize the dictionary objective function. The sparse coding property allows a kernel with a compact support in GP to realize a very efficient dictionary learning process. Hence, video of an activity can be described by a set of compact and discriminative action attributes.

Given the initial dictionary $\mathbf{B}^o$, the objective is to compress it into a dictionary $\mathbf{B}^*$ of size k, which encourages the signals from the same class to have very similar sparse representations. Let L denote the labels of M discrete values, $L \in [1, M]$. Given a set of dictionary atoms $\mathbf{B}^*$, define $P(L|\mathbf{B}^*) = \frac{1}{|\mathbf{B}^*|} \Sigma_{\mathbf{b}_i \in \mathbf{B}^*} P(L|\mathbf{b}_i)$. For simplicity, denote $P(L|\mathbf{b}^*)$ as $P(L_{\mathbf{b}^*})$, and $P(L|\mathbf{B}^*)$ as $P(L_{\mathbf{B}^*})$. To enhance the discriminative power of the learned dictionary, the following objective function is considered

$$\arg\max_{\mathbf{B}^*} I(\mathbf{B}^*; \mathbf{B}^o \setminus \mathbf{B}^*) + \lambda I(L_{\mathbf{B}^*}; L_{\mathbf{B}^o \setminus \mathbf{B}^*}) \qquad (6.2)$$

where $\lambda \geq 0$ is the parameter to regularize the emphasis on appearance or label information and I denotes mutual information. One can approximate (6.2) as

$$\arg\max_{\mathbf{b}^* \in \mathbf{B}^o \setminus \mathbf{B}^*} [H(\mathbf{b}^*|\mathbf{B}^*) - H(\mathbf{b}^*|\bar{\mathbf{B}}^*)]$$

$$+ \lambda [H(L_{\mathbf{b}^*}|L_{\mathbf{B}^*}) - H(L_{\mathbf{b}^*}|L_{\bar{\mathbf{B}}^*})], \qquad (6.3)$$

where H denotes entropy. One can easily notice that the above formulation also forces the classes associated with $\mathbf{b}^*$ to be most different from classes already covered by the selected atoms $\mathbf{B}^*$; and at the same time, the classes associated with $\mathbf{b}^*$ are most representative among classes covered by the remaining atoms. Thus the learned dictionary is not only compact, but also covers all classes to maintain the discriminability.

In Figure 6.1, we present the recognition accuracy on the Keck gesture dataset with different dictionary sizes and over different global and local features [115]. Leave-one-person-out setup is used. That is, sequences performed by a person are left out, and the average accuracy is reported. Initial dictionary size $|\mathbf{B}^o|$ is chosen to be twice the dimension of the input signal and sparsity 10 is used in this set of experiments. As can be seen the mutual information-based method, denoted as MMI-2 outperforms the other methods.

Sparse representation over a dictionary with coherent atoms has the multiple representation problem. A compact dictionary consists of incoherent atoms, and encourages similar signals, which are more likely from the same class, to be consistently described by a similar set of atoms with similar coefficients [115]. A discriminative dictionary encourages signals from different classes to be described by either a different set of atoms, or the same set of atoms but with different coefficients [71, 82, 119]. Both aspects are critical for classification using sparse representation. The reconstructive requirement to a compact and discriminative dictionary enhances the robustness of the discriminant sparse representation [119].

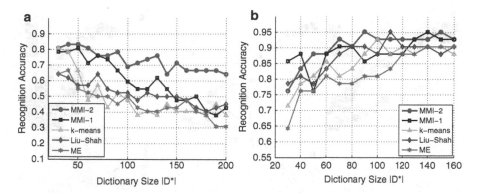

Fig. 6.1 Recognition accuracy on the Keck gesture dataset with different features and dictionary sizes (shape and motion are global features. STIP is a local feature.) [115]. The recognition accuracy using initial dictionary D^o: (a) 0.23 (b) 0.42. In all cases, the MMI-2 (red line) outperforms the rest

Hence, learning reconstructive, compact and discriminative dictionaries are important for classification using sparse representation. Motivated by this observation, Qiu *et al.* [114] proposed a general information theoretic approach to leaning dictionaries that are simultaneously reconstructive, discriminative and compact.

Suppose that we are given a set of n signals (images) in an N-dim feature space $\mathbf{X} = [\mathbf{x_1}, ..., \mathbf{x_N}]$, $\mathbf{x_i} \in \mathbb{R}^N$. Given that signals are from p distinct classes and N_c signals are from the c-th class, $c \in \{1, \cdots, p\}$, we denote $\Gamma = \{\Gamma_c\}_{c=1}^p$, where $\Gamma_c = [\gamma_1^c, \cdots, \gamma_{N_c}^c]$ are signals in the c-th class. Define $\Gamma = \{\Gamma_c\}_{c=1}^p$, where $\Gamma_c = [\gamma_1^c, \cdots, \gamma_{N_c}^c]$ is the sparse representation of $\mathbf{X}_c$.

Given $\mathbf{X}$ and an initial dictionary $\mathbf{B}^o$ with ℓ_2 normalized columns, a compact, reconstructive and discriminative dictionary $\mathbf{B}^*$ is learned via maximizing the mutual information between $\mathbf{B}^*$ and the unselected atoms $\mathbf{B}^o \backslash \mathbf{B}^*$ in $\mathbf{B}^o$, between the sparse codes $\Gamma_{\mathbf{B}^*}$ associated with $\mathbf{B}^*$ and the signal class labels C, and finally between the signals $\mathbf{Y}$ and $\mathbf{D}^*$, i.e.,

$$\arg\max_{\mathbf{B}} \lambda_1 I(\mathbf{B}; \mathbf{B}^o \backslash \mathbf{B}) + \lambda_2 I(\Gamma_{\mathbf{B}}; C) + \lambda_3 I(\mathbf{X}; \mathbf{B}) \qquad (6.4)$$

where $\{\lambda_1, \lambda_2, \lambda_3\}$ are the parameters to balance the contributions from compactness, discriminability and reconstruction terms, respectively.

A two-stage approach is adopted to satisfy (6.4). In the first stage, each term in (6.4) is maximized in a unified greedy manner and involves a closed-form evaluation, thus atoms can be greedily selected from the initial dictionary while satisfying (6.4). In the second stage, the selected dictionary atoms are updated using a simple gradient ascent method to further maximize

$$\lambda_2 I(\Gamma_{\mathbf{B}}; C) + \lambda_3 I(\mathbf{X}; \mathbf{B}).$$

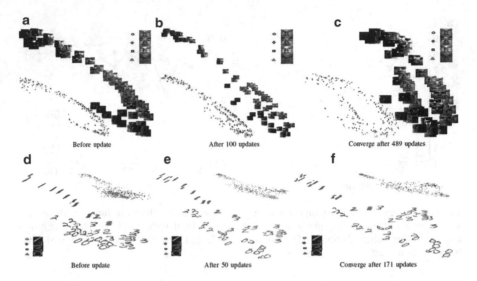

Fig. 6.2 Information-theoretic dictionary update with global atoms shared over classes. For a better visual representation, sparsity 2 is chosen and a randomly selected subset of all samples are shown. The recognition rate associated with (a), (b), and (c) are: 30.63%, 42.34% and 51.35%. The recognition rate associated with (d), (e), and (f) are: 73.54%, 84.45% and 87.75%. Note that the ITDU effectively enhances the discriminability of the set of common atoms [114]

To illustrate how the discriminability of dictionary atoms selected by the information theoretic dictionary section (ITDS) method can be further enhanced using the information theoretic dictionary update (ITDU) method, consider Fig. 6.2. The Extended YaleB face dataset [64] and the USPS handwritten digits dataset [1] are used for illustration. Sparsity 2 is adopted for visualization, as the non-zero sparse coefficients of each image can now be plotted as a 2-D point. In Fig. 6.2, with a common set of atoms shared over all classes, sparse coefficients of all samples become points in the same 2-D coordinate space. Different classes are represented by different colors. The original images are also shown and placed at the coordinates defined by their non-zero sparse coefficients. The atoms to be updated in Fig. 6.2(a) and 6.2(d) are selected using ITDS. One can see from Fig. 6.2 that the ITDU method makes sparse coefficients of different classes more discriminative, leading to significantly improved classification accuracy [114].

6.3   Non-Linear Kernel Dictionary Learning

Similar to finding non-linear sparse representation in the high dimensional feature space, one can also lean non-linear dictionaries using the kernel methods. Let $\Phi :$ $\mathbb{R}^N \to \mathscr{F} \subset \mathbb{R}^{\tilde{N}}$ be a non-linear mapping from $\mathbb{R}^N$ into a dot product space $\mathscr{F}$. One can learn a non-linear dictionary $\mathbf{B}$ in the feature space $\mathscr{F}$ by solving the following optimization problem:

$$\underset{\mathbf{B},\Gamma}{\arg\min} \ \|\Phi(\mathbf{X}) - \mathbf{B}\Gamma\|_F^2 \ s.t \ \|\gamma_i\|_0 \leq T_0, \forall i. \tag{6.5}$$

where $\mathbf{B} \in \mathbb{R}^{\tilde{N} \times K}$ is the sought dictionary, $\Gamma \in \mathbb{R}^{K \times n}$ is a matrix whose ith column is the sparse vector γ_i corresponding to the sample $\mathbf{x}_i$, with maximum of T_0 non-zero entries. It was shown in [95], that there exists an optimal solution $\mathbf{B}^*$ to the problem (6.5) that has the following form:

$$\mathbf{B}^* = \Phi(\mathbf{X})\mathbf{A} \tag{6.6}$$

for some $\mathbf{A} \in \mathbb{R}^{n \times K}$. Moreover, this solution has the smallest Frobenius norm among all optimal solutions. As a result, one can seek an optimal dictionary through optimizing $\mathbf{A}$ instead of $\mathbf{B}$. By substituting (6.6) into (6.5), the problem can be re-written as follows:

$$\underset{\mathbf{A},\Gamma}{\arg\min} \ \|\Phi(\mathbf{X}) - \Phi(\mathbf{X})\mathbf{A}\Gamma\|_F^2 \ s.t \ \|\gamma_i\|_0 \leq T_0, \forall i. \tag{6.7}$$

In order to see the advantage of this formulation over the original one, we will examine the objective function. Through some manipulation, the cost function can be re-written as:

$$\|\Phi(\mathbf{X}) - \Phi(\mathbf{X})\mathbf{A}\Gamma\|_F^2 = \mathrm{tr}((\mathbf{I} - \mathbf{A}\Gamma)^T \mathbb{K}(\mathbf{X},\mathbf{X})(\mathbf{I} - \mathbf{A}\Gamma)),$$

where $\mathbb{K}(\mathbf{X},\mathbf{X})$ is a kernel matrix whose elements are computed from $\kappa(i,j) = \Phi(\mathbf{x}_i)^T \Phi(\mathbf{x}_j)$. It is apparent that the objective function is feasible since it only involves a matrix of finite dimension $\mathbb{K} \in \mathbb{R}^{n \times n}$, instead of dealing with a possibly infinite dimensional dictionary. An important property of this formulation is that the computation of $\mathbb{K}$ only requires dot products. Therefore, we are able to employ *Mercer* kernel functions to compute these dot products without carrying out the mapping Φ.

To solve the above optimization problem for learning non-linear dictionaries, variants of MOD and K-SVD algorithms in the feature space have been proposed [96], [95]. The procedure essentially involves two stages: sparse coding and dictionary update in the feature space. For sparse coding, one can adapt the non-linear version of orthogonal matching pursuit algorithm [95]. Once the sparse codes are found in the feature space, the dictionary atoms are updated in an efficient way [96], [95].

The optimization problem (6.7) is purely generative. It does not explicitly promote the discrimination which is important for many classification tasks. Using the kernel trick, when the data is transformed into a high dimensional feature space, the data from different classes may still overlap. Hence, generative dictionaries may lead to poor performance in classification even when data is non-linearly mapped to a feature space. To overcome this, a method for designing non-linear dictionaries which are simultaneously generative and discriminative was proposed in [132].

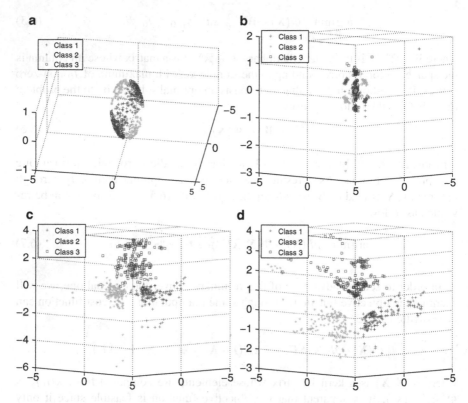

Fig. 6.3 A synthetic example showing the significance of learning a discriminative dictionary in feature space for classification. (a) Synthetic data which consists of linearly non separable 3D points on a sphere. Different classes are represented by different colors. (b) Sparse coefficients from K-SVD projected onto learned SVM hyperplanes. (c) Sparse coefficients from a non-linear dictionary projected onto learned SVM hyperplanes. (d) Sparse coefficients from non-linear discriminative kernel dictionary projected onto learned SVM hyperplanes [132]

Figure 6.3 presents an important comparison in terms of the discriminative power of learning a discriminative dictionary in the feature space where kernel LDA type of discriminative term has been included in the objective function. A scatter plot of the sparse coefficients obtained using different approaches show that such a discriminative dictionary is able to learn the underlying non-linear sparsity of data as well as it provides more discriminative representation. See [132], [96], [95] for more details on the design of non-linear kernel dictionaries.

Chapter 7
Concluding Remarks

In this monograph, we have outlined some fundamental premises underlying the theories of sparse representation and compressed sensing. Based on these theories, we have examined several interesting imaging and vision applications such as magnetic resonance imaging, synthetic aperture radar imaging, millimeter wave imaging, target tracking, background subtraction, video processing and biometrics recognition.

Although significant progress has been made, a number of challenges and issues confront the effectiveness of sparse representations and compressive sampling in imaging and computer vision problems. Below we list a few.

- *Sparse representation-based recognition from video*: [112] attempts to propose a method for iris recognition from video based on sparse representation. Is it possible to extend this method for other biometrics recognition from video? In this case the columns of the dictionary are composed of "dynamic features" extracted from the video samples of each subject in the dataset. A video-based reduction of redundancy may be easily applied to reduce the dimensionality of the dictionary matrix.
- *Multi-modal object recognition based on sparse representation and CS*: The topic of multi-modal biometrics has gained strong interest in recent years [120]. In this approach, multiple biometrics data (either coming from the same sensing device or from different sources) are fused together and processed with a single matching algorithm or with several concurrent algorithms. The scores produced by different algorithms can be also fused to produce a single matching score for identification. Can sparse representation and CS-based methods offer better solutions for multimodal biometrics fusion?
- *Non-linear kernel dictionary learning*: Uniqueness and stability of the learned kernel dictionary has not been established. More theoretical work is needed in the area of non-linear dictionary learning..
- *Feature extraction from compressed measurements*: Extraction of salient features is one of the main problems in computer vision. Can we develop features that contain the same information about the object in the compressed measurements?

V.M. Patel and R. Chellappa, *Sparse Representations and Compressive Sensing for Imaging and Vision*, SpringerBriefs in Electrical and Computer Engineering, DOI 10.1007/978-1-4614-6381-8_7, © The Author(s) 2013

- *Number of training samples*: The methods presented in [156] and [112] harnessing sparsity are very effective yet they suffer from some limitations. For instance, for good recognition performance, the training image set is required to be extensive enough to span the conditions that might occur in the test set. For example in the case of face biometric, to be able to handle illumination variations in the test image, more and more training images are needed in the gallery. But in most realistic scenarios, the gallery contains only a single or a few images of each subject and it is not practical to assume the availability of multiple images of the same person under different illumination conditions. Another limitation of this approach is that the large size of the matrix, due to the inclusion of the large number of gallery images, can tremendously increase the computational as well as the storage complexity which can make the real-time processing very difficult. Can sparsity motivated dictionary learning methods offer solution to this problem?

References

1. Usps handwritten digit database. In *http://www-i6.informatik.rwth-aachen.de/keysers/usps.html*.
2. M. Aharon, M. Elad, and A. M. Bruckstein. The k-svd: an algorithm for designing of overcomplete dictionaries for sparse representation. *IEEE Trans. Signal Process.*, 54(11):4311–4322, 2006.
3. F. Alizadeh and D. Goldfarb. Second-order cone programming. *Mathematics Programming*, 95(1):3–51, 2003.
4. R. Appleby and R. N. Anderton. Millimeter-wave and submillimeter-wave imaging for security and surveillance. *Proceedings of the IEEE*, 95(8):1683 –1690, Aug. 2007.
5. S. D. Babacan, M. Luessi, L. Spinoulas, and A. K. Katsaggelos. Compressive passive millimeter-wave imaging. In *IEEE ICIP*, 2011.
6. G. R. Baraniuk. Compressive sensing. *IEEE Signal Processing Magazine*, 24(4):118–121, July 2007.
7. R. Baraniuk. More is less: signal processing and the data deluge. *Science*, 331(6018):717–719, 2011.
8. E. van den Berg and M. P. Friedlander. Probing the Pareto frontier for basis pursuit solutions. Technical report, Department of Computer Science, University of British Columbia.
9. J. M. Bioucas-Dias and M. A. T. Figueiredo. A New TwIST: Two-Step Iterative Shrinkage/ Thresholding Algorithms for image restoration. *IEEE Transactions on Image Processing*, 16(12):2992–3004, Dec. 2007.
10. J. D. Blanchard, C. Cartis, J. Tanner, and A. Thompson. Phase transitions for greedy sparse approximation algorithms. *Applied and Computational Harmonic Analysis*, 30(2):188–203, 2011.
11. T. Blumensath and E. Davies. Iterative hard thresholding for compressed sensing. *preprint*.
12. T. Blumensath and E. Davies. Iterative thresholding for sparse approximations. *The Journal of Fourier Analysis and Applications*, 14(5):629–654, Dec. 2008.
13. T. Blumensath and M. E Davies. Signal recovery from partial information via orthogonal matching pursuit. *IEEE Transactions on Signal Processing*, 56(6):2370–2382, June 2008.
14. K. W. Bowyer and P. J. Flynn. The nd-iris-0405 iris image dataset. *Notre Dame CVRL Technical Report*.
15. S. Boyd and L. Vandenberghe. *Convex Optimization*. Cambridge University Press, New York, NY, 2004.
16. Sara Bradburn, Wade Thomas Cathey, and Edward R. Dowski. Realizations of focus invariance in optical–digital systems with wave-front coding. *Appl. Opt.*, 36(35):9157–9166, Dec. 1997.

17. Alfred M. Bruckstein, David L. Donoho, and Michael Elad. From sparse solutions of systems of equations to sparse modeling of signals and images. *SIAM Rev.*, 51(1):34–81, February 2009.
18. Christopher J.C. Burges. A tutorial on support vector machines for pattern recognition. *Data Mining and Knowledge Discovery*, 2:121–167, 1998.
19. E. Candes and J. Romberg. Practical signal recovery from random projections. *Proceedings of the SPIE*, 5674:76–86, 2005.
20. E. Candes and J. Romberg. Signal recovery from random projections. *in Proc. of SPIE Computational Imaging III*, 5674, 2005.
21. E. Candes and J. Romberg. Quantitatively robust uncertainty principles and optimally sparse decompositions. *Foundations of Comput. Math.*, 6(2):227–254, April 2006.
22. E. Candes, J. Romberg, and T. Tao. Stable signal recovery from incomplete and inaccurate measurements. *Communications on Pure and Applied Mathematics*, 59(8):1207–1223, August 2006.
23. E. Candes, J. Romberg, and T. Tao. Robust uncertainty principles: exact signal reconstruction from highly incomplete frequency information. *IEEE Transactions on Information Theory*, 52(2):489–509, Feb. 2006.
24. E. Candes and T. Tao. Decoding by linear programing. *IEEE Transactions on Information Theory*, 51(12):4203–4215, Dec. 2005.
25. E. Candes and T. Tao. Near optimal signal recovery from random projections: Universal encoding strategies? *IEEE Transactions on Information Theory*, 52(12):5406–5425, Dec. 2006.
26. E. J. Candes. Compressive sampling. *International Congress of Mathematics, Madrid, Spain*, 3:1433–1452, 2006.
27. E. J. Candes, X. Li, Y. Ma, and J. Wright. Robust principal component analysis? *Journal of ACM*, 58(1):1–37, 2009.
28. E. J. Candes and T Tao. The Dantzig selector: statistical estimation when p is much larger than n. *Annals of Statistics*, 35(6):2313–2351, 2007.
29. E. J. Candes, M. Wakin, and S. Boyd. Enhancing sparsity by reweighted ℓ_1 minimization. *J. Fourier Anal. Appl.*, 14:877–905, 2008.
30. W. G. Carrara, R. S. Goodman, and R. M. Majewski. *Spotlight Synthetic Aperture Radar: Signal Processing Algorithms*. Artech House, Norwood, MA, 1995.
31. W. Thomas Cathey and Edward R. Dowski. New paradigm for imaging systems. *Appl. Opt.*, 41(29):6080–6092, Oct. 2002.
32. M. Çetin and W. C. Karl. Feature-enhanced synthetic aperture radar image formation based on nonquadratic regularization. *IEEE Transactions on Image Processing*, 10(4):623–631, Apr. 2001.
33. V. Cevher, A. Sankaranarayanan, M. Duarte, D. Reddy, R. Baraniuk, and R. Chellappa. Compressive sensing for background subtraction. *ECCV*, 2008.
34. A. B. Chan and N. Vasconcelos. Probabilistic kernels for the classification of auto-regressive visual processes. In *IEEE Conf. on Computer Vision and Pattern Recognition*, pages 846–851, 2005.
35. T. F. Chan, S. Esedoglu, F. Park, and M. H. Yip. *Recent Developments in Total Variation Image Restoration*. Springer Verlag, 2005.
36. Wai Lam Chan, Kriti Charan, Dharmpal Takhar, Kevin F. Kelly, Richard G. Baraniuk, and Daniel M. Mittleman. A single-pixel terahertz imaging system based on compressed sensing. *Appl. Phys. Lett.*, 93(12):121105–3, 2008.
37. R. Chartrand. Exact reconstructions of sparse signals via nonconvex minimization. *IEEE Signal Processing Letters*, 14:707–710, 2007.
38. S. Chen, D. Donoho, and M. Saunders. Atomic decomposition by basis pursuit. *SIAM J. Sci. Comp.*, 20(1):33–61, 1998.
39. V. C. Chen and H. Ling. *Time-Frequency Transforms for Radar Imaging and Signal Analysis*. Artech House, Norwood, MA, 2002.
40. Y-C. Chen, V. M. Patel, P. J. Phillips, and R. Chellappa. Dictionary-based face recognition from video. In *European Conference on Computer Vision*, pages 1–14, 2012.

41. Yi Chen, N.M. Nasrabadi, and T.D. Tran. Hyperspectral image classification via kernel sparse representation. In *IEEE International Conference on Image Processing*, pages 1233–1236, sept. 2011.
42. M. Cossalter, G. Valenzise, M. Tagliasacchi, and S. Tubaro. Joint compressive video coding and analysis. *IEEE Transactions on Multimedia*, 12(3):168–183, 2010.
43. Christy Fernandez Cull, David A. Wikner, Joseph N. Mait, Michael Mattheiss, and David J. Brady. Millimeter-wave compressive holography. *Appl. Opt.*, 49(19):E67–E82, Jul 2010.
44. I. G. Cumming and F. H. Wong. *Digital processing of synthetic aperture radar data*. Artech House, Norwood, MA, 2005.
45. I. Daubechies, M. Defries, and C. De Mol. An iterative thresholding algorithm for linear inverse problems with a sparsity constraint. *Communications on Pure and Applied Mathematics*, 57:1413–1457, 2004.
46. J. Daugman. How iris recognition works. *IEEE Transactions on Circuits and Systems for Video Technology*, 14(1):21–30, Jan. 2004.
47. D. Donoho. Compressed sensing. *IEEE Transactions on Information Theory*, 52(4):1289–1306, Apr. 2006.
48. D. L. Donoho. High-dimensional centrally symmetric polytopes with neighborliness proportional to dimension. *Discrete and Comput. Geom.*, 35(4):617652, 2006.
49. D. L. Donoho, Y. Tsaig, I. Drori, and J-L. Starck. Sparse solution of underdetermined linear equations by Stagewise Orthogonal Matching Pursuit. *IEEE Transactions on Information Theory*, March 2006, preprint.
50. David L. Donoho and Michael Elad. Optimally sparse representation in general (non-orthogonal) dictionaries via l1 minimization. *Proc. Natl Acad. Sci.*, 100:2197–2202, 2003.
51. D.L. Donoho and J. Tanner. Precise undersampling theorems. *Proceedings of the IEEE*, 98(6):913–924, june 2010.
52. L. Donoho, D. and Y. Tsaig. Recent advances in sparsity-driven signal recovery. *Proceedings of IEEE International Conference on Acoustics, Speech and Signal Processing*, 5:v713–v716, March 2005.
53. G. Doretto, A. Chiuso, Y.N. Wu, and S. Soatto. Dynamic textures. *International Journal of Computer Vision*, 51(2):91–109, 2003.
54. M.F. Duarte, M.A. Davenport, D. Takhar, J.N. Laska, Ting Sun, K.F. Kelly, and R.G. Baraniuk. Single-pixel imaging via compressive sampling. *IEEE Signal Processing Magazine*, 25(2):83–91, march 2008.
55. M.F. Duarte and Y.C. Eldar. Structured compressed sensing: From theory to applications. *IEEE Transactions on Signal Processing,*, 59(9):4053–4085, sept. 2011.
56. Jr. Edward R. Dowski and W. Thomas Cathey. Extended depth of field through wave-front coding. *Appl. Opt.*, 34(11):1859–1866, Apr. 1995.
57. M. Elad. *Sparse and Redundant Representations: From theory to applications in Signal and Image processing*. Springer, 2010.
58. K. Engan, S. O. Aase, and J. H. Husoy. Method of optimal directions for frame design. *Proc. IEEE Int. Conf. Acoust., Speech, Signal Process.*, 5:2443–2446, 1999.
59. K. Etemand and R. Chellappa. Separability-based multiscale basis selection and feature extraction for signal and image classification. *IEEE Transactions on Image Processing*, 7(10):1453–1465, Oct. 1998.
60. Christy A. Fernandez, David Brady, Joseph N. Mait, and David A. Wikner. Sparse fourier sampling in millimeter-wave compressive holography. In *Digital Holography and Three-Dimensional Imaging*, page JMA14, 2010.
61. M. A. T. Figueiredo, R. D. Nowak, and S. J. Wright. Gradient Projection for Sparse Reconstruction: Application to compressed sensing and other inverse problems. *IEEE Journal of Selected Topics in Signal Processing*, 1(4):586–598, 2007.
62. Robert T. Frankot and Rama Chellappa. A method for enforcing integrability in shape from shading algorithms. *IEEE Trans. Pattern Anal. Mach. Intell.*, 10(4):439–451, 1988.
63. S. Gao, I. W. Tsang, and L.-T. Chia. Kernel sparse representation for image classification and face recognition. In *European Conference on Computer Vision*, volume 6314, 2010.

64. A. S. Georghiades, P. N. Belhumeur, and D. J. Kriegman. From few to many: Ilumination cone models for face recognition under variable lighting and pose. *IEEE Trans. Pattern Analysis and Machine Intelligence*, 23(6):643–660, June 2001.

65. T. Goldstein and S. Osher. The split Bregman algorithm for L1 regularized problems. Technical report, UCLA CAM.

66. J. W. Goodman. *Introduction to Fourier optics*. Englewood, CO: Roberts and Company, 2005.

67. N. Gopalsami, T. W. Elmer, S. Liao, R. Ahern, A. Heifetz, A. C. Raptis, M. Luessi, S. D. Babacan, and A. K. Katsaggelos. Compressive sampling in passive millimeter-wave imaging. In *Proc. SPIE*, volume 8022, 2011.

68. I. F. Gorodnitsky and B. D. Rao. Sparse signal reconstruction from limited data using FOCUSS: A re-weighted minimum norm algorithm. *IEEE Transactions on Signal Processing*, 45(3):600616, March 1997.

69. R. Gribonval and M. Nielsen. Sparse representations in unions of bases. *IEEE Transactions on Information Theory*, 49(12):3320 – 3325, dec. 2003.

70. W. Guoa and W. Yin. Edgecs: Edge guided compressive sensing reconstruction. *Rice University CAAM Technical Report TR10-02*, 2010.

71. K. Huang and S. Aviyente. Sparse representation for signal classification. *NIPS*, 19:609–616, 2007.

72. A.K. Jain, S. Prabhakar, L. Hong, and S. Pankanti. Filterbank-based fingerprint matching. *IEEE Transactions on Image Processing*, 9(5):846–859, May 2000.

73. Zhuolin Jiang, Zhe Lin, and L.S. Davis. Learning a discriminative dictionary for sparse coding via label consistent k-svd. In *Computer Vision and Pattern Recognition (CVPR), 2011 IEEE Conference on*, pages 1697 –1704, june 2011.

74. S-J. Kim, K. Koh, M. Lustig, S. Boyd, and D. Gorinevsky. An interior-point method for large-scale ℓ_1-regularized least squares. *IEEE Journal of Selected Topics in Signal Processing*, 1(4):606–617, Dec. 2007.

75. E. Kokiopoulou and P. Frossard. Semantic coding by supervised dimensionality reduction. *IEEE Trans. Multimedia*, 10(5):806–818, Aug. 2008.

76. P. Kovesi. Shapelets correlated with surface normals produce surfaces. In *ICCV*, volume 2, pages 994 –1001 Vol. 2, oct. 2005.

77. B. Krishnapuram, L. Carin, M.A.T. Figueiredo, and A.J. Hartemink. Sparse multinomial logistic regression: fast algorithms and generalization bounds. *IEEE Transactions on Pattern Analysis and Machine Intelligence*, 27(6):957 –968, June 2005.

78. S. Lazebnik, C. Schmid, and J. Ponce. Beyond bags of features: Spatial pyramid matching for recognizing natural scene categories. In *Computer Vision and Pattern Recognition, 2006 IEEE Computer Society Conference on*, volume 2, pages 2169 – 2178, 2006.

79. M. Lustig, D. Donoho, and M. J. Pauly. Sparse MRI: The Application of Compressed Sensing for Rapid MR Imaging. *Magnetic Resonance in Medicine*, 58(6):1182–1195, Dec. 2007.

80. M. Lustig, D.L. Donoho, J.M. Santos, and J.M. Pauly. Compressed sensing mri. *IEEE Signal Processing Magazine*, 25(2):72 –82, march 2008.

81. X. Feng M. Yang, L. Zhang and D. Zhang. Fisher discrimination dictionary learning for sparse representation, 2011. *ICCV*.

82. J. Mairal, F. Bach, J. Pnce, G. Sapiro, and A. Zisserman. Discriminative learned dictionaries for local image analysis. *Proc. of the Conference on Computer Vision and Pattern Recognition*, 2008.

83. J. Mairal, F. Bach, and J. Ponce. Task-driven dictionary learning. *IEEE Transactions on Pattern Analysis and Machine Intelligence*, 34(4):791 –804, april 2012.

84. J. Mairal, F. Bach, J. Ponce, G. Sapiro, and A. Zisserman. Supervised dictionary learning. *Advances in Neural Information Processing Systems*, 2008.

85. J.N. Mait, D.A. Wikner, M.S. Mirotznik, J. van der Gracht, G.P. Behrmann, B.L. Good, and S.A. Mathews. 94-GHz imager with extended depth of field. *IEEE Transactions on Antennas and Propagation*, 57(6):1713 –1719, june 2009.

86. A. Maleki and D.L. Donoho. Optimally tuned iterative reconstruction algorithms for compressed sensing. *IEEE Journal of Selected Topics in Signal Processing*, 4(2):330–341, april 2010.

87. S Mallat. *A wavelet tour of signal processing, second edition*. Academic Press, San Diego, CA, 1999.
88. S. Mallat and Z. Zhang. Matching Pursuit with time-frequency dictionaries. *IEEE Transactions on Signal Processing*, 41(12):33973415, 1993.
89. L. Masek and P. Kovesi. Matlab source code for a biometric identification system based on iris patterns. *The School of Computer Science and Software Engineering, The University of Western Australia*, 2003.
90. L. Meier, S. V. D. Geer, and P. Bhlmann. The group lasso for logistic regression. *Journal of the Royal Statistical Society: Series B*, 70(1):53–71, 2008.
91. L. B. Montefusco, D. Lazzaro, and S. Papi. Nonlinear filtering for sparse signal recovery from incomplete measurements. *IEEE Trans. Sig. Proc.*, 57(7):2494–2502, July 2009.
92. Jr. Munson, D.C., J.D. O'Brien, and W.K. Jenkins. A tomographic formulation of spotlight-mode synthetic aperture radar. *Proceedings of the IEEE*, 71(8):917 – 925, aug. 1983.
93. D. Needell and J. A. Tropp. Cosamp: Iterative signal recovery from incomplete and inaccurate samples. *Appl. Comp. Harmonic Anal.*, 26:301–321, 2008.
94. D. Needell and R. Vershynin. Signal recovery from incomplete and inaccurate measurements via regularized orthogonal matching pursuit. *IEEE Journal of Selected Topics in Signal Processing*, 4(2):310 –316, april 2010.
95. H. V. Nguyen, V. M. Patel, N. M. Nasrabadi, and R. Chellappa. Design of non-linear kernel dictionaries for object recognition. *IEEE Transactions on Pattern Analysis and Machine Intelligence*, submitted 2012.
96. H. V. Nguyen, V. M. Patel, N. M. Nasrabadi, and R. Chellappa. Kernel dictionary learning. In *IEEE International Conference on Acoustics, Speech and Signal Processing*, 2012.
97. Nam H. Nguyen, Nasser M. Nasrabadi, and Trac D. Tran. Robust multi-sensor classification via joint sparse representation. In *International Conference on Information Fusion*, 2011.
98. Imoma Noor, Orges Furxhi, and Eddie L. Jacobs. Compressive sensing for a sub-millimeter wave single pixel imager. In *Proc. SPIE*, volume 8022, 2011.
99. B. A. Olshausen and D. J. Field. Emergence of simple-cell receptive field properties by learning a sparse code for natural images. *Nature*, 381(6583):607–609, 1996.
100. V. M. Patel and R. Chellappa. Sparse representations, compressive sensing and dictionaries for pattern recognition. In *Asian Conference on Pattern Recognition*, 2010.
101. V. M. Patel, G. R. Easley, and D. M. Healy. Automatic target recognition based on simultaneous sparse representation. *IEEE International Conference on Image Processing*, submitted, 2010.
102. V. M. Patel, G. R. Easley, D. M. Healy, and R. Chellappa. Compressed sensing for synthetic aperture radar imaging. *International Conference on Image Processing, submitted*, 2009.
103. V. M. Patel, G. R. Easley, D. M. Healy, and R. Chellappa. Compressed synthetic aperture radar. *IEEE Journal of Selected Topics in Signal Processing*, 4(2):244–254, April 2010.
104. V. M. Patel and J. N. Mait. Compressive passive millimeter-wave imaging with extended depth of field. *Optical Engineering*, 51(9), 2012.
105. V. M. Patel and J. N. Mait. Passive millimeter-wave imaging with extended depth of field and sparse data. *IEEE International Conference on Acoustics, Speech and Signal Processing*, 2012.
106. V. M. Patel, N. M. Nasrabadi, and R. Chellappa. Sparsity-motivated automatic target recognition. *Applied Optics*, 50(10), April 2011.
107. Vishal M. Patel, Tao Wu, Soma Biswas, P. Jonathon Phillips, and Rama Chellappa. Dictionary-based face recognition under variable lighting and pose. *IEEE Transactions on Information Forensics and Security*, 7(3):954–965, 2012.
108. V.M. Patel, R. Maleh, A.C. Gilbert, and R. Chellappa. Gradient-based image recovery methods from incomplete fourier measurements. *IEEE Transactions on Image Processing*, 21(1):94 –105, jan. 2012.
109. Y. C. Pati, R. Rezaiifar, and P. S. Krishnaprasad. Orthogonal Matching Pursuit: Recursive function approximation with applications to wavelet decomposition. *In Proc. 27th Ann. Asilomar Conf. Signals, Systems, and Computers*, Nov. 1993.

110. P. Peers, D. K. Mahajan, B. Lamond, A. Ghosh, W. Matusik, R. Ramamoorthi, and P. Debevec. Compressive light transport sensing. *ACM Trans. Graph.*, 28(1):3:1–3:18, February 2009.
111. J. Pillai, V. M. Patel, and R. Chellappa. Sparsity inspired selection and recognition of iris images. *Third IEEE International Conference on Biometrics : Theory, Applications and Systems*, 2009.
112. Jaishanker K. Pillai, Vishal M. Patel, Rama Chellappa, and Nalini K. Ratha. Secure and robust iris recognition using random projections and sparse representations. *IEEE Trans. Pattern Anal. Mach. Intell.*, 33(9):1877–1893, September 2011.
113. S.J. Pundlik, D.L. Woodard, and S.T. Birchfield. Non-ideal iris segmentation using graph cuts. In *IEEE CVPR Workshop*, pages 1–6, June 2008.
114. Q. Qiu, V. M. Patel, and R. Chellappa. Information-theoretic dictionary learning for image classification. *IEEE Transactions on Pattern Analysis and Machine Intelligence*, submitted 2012.
115. Qiang Qiu, Zhuolin Jiang, and Rama Chellappa. Sparse dictionary-based representation and recognition of action attributes. In *International Conference on Computer Vision*, 2011.
116. D. Reddy, A. Sankaranarayanan, V. Cevher, and R. Chellappa. Compressed sensing for multi-view tracking and 3-d voxel reconstruction. *IEEE International Conference on Image Processing*, (4):221–224, 2008.
117. D. Reddy, A. Veeraraghavan, and R. Chellappa. P2c2: Programmable pixel compressive camera for high speed imaging. *Computer Vision and Pattern Recognition, IEEE Computer Society Conference on*, pages 329–336, 2011.
118. Dikpal Reddy, Amit K. Agrawal, and Rama Chellappa. Enforcing integrability by error correction using l1-minimization. In *CVPR*, pages 2350–2357, 2009.
119. F. Rodriguez and G. Sapiro. Sparse representations for image classification: Learning discriminative and reconstructive non-parametric dictionaries. *Tech. Report, University of Minnesota*, Dec. 2007.
120. A. Ross, K. Nandakumar, and A. K. Jain. *Handbook of Multibiometrics*. Springer, 2006.
121. R. Rubinstein, A.M. Bruckstein, and M. Elad. Dictionaries for sparse representation modeling. *Proceedings of the IEEE*, 98(6):1045 –1057, june 2010.
122. M. Rudelson and R. Vershynin. On sparse reconstruction from fourier and gaussian measurements. *Comm. on Pure and Applied Mathematics*, 61:1025–1045, 2008.
123. C. Wu S. Chikkerur and V. Govindaraju. A systematic approach for feature extraction in fingerprint images. In *Int. Conference on Bioinformatics and its Applications*, page 344, 2004.
124. S. Schuckers S. Crihalmeanu, A. Ross and L. Hornak. A protocol for multibiometric data acquisition, storage and dissemination. In *Technical Report, WVU, Lane Department of Computer Science and Electrical Engineering*, 2007.
125. A. Sankaranarayanan, , A. Veeraraghavan, and R. Chellappa. Object detection, tracking and recognition for multiple smart cameras. *Proceedings of the IEEE*, 96(10):1606–1624, 2008.
126. A. Sankaranarayanan and R. Chellappa. Optimal multi-view fusion of object locations. *IEEE Workshop on Motion and Video Computing*, pages 1–8, 2008.
127. Aswin C. Sankaranarayanan, Pavan K. Turaga, Richard G. Baraniuk, and Rama Chellappa. Compressive acquisition of dynamic scenes. In *ECCV*, pages 129–142, 2010.
128. B. Scholkopf and A. J. Smola. *Learning With Kernels, Support Vector Machines, Regularization, Optimization, and Beyond*. MIT Press, 2001.
129. H. S. Shapiro and R. A. Silverman. Alias-free sampling of random noise. *SIAM*, 8(2):225–248, June 1959.
130. S. Shekhar, V. M. Patel, N. M. Nasrabadi, and R. Chellappa. Joint sparsity-based robust multimodal biometrics recognition. *IEEE Transactions on Pattern Analysis and Machine Intelligence*, to be submitted 2012.
131. S. Shekhar, V.M. Patel, and R. Chellappa. Synthesis-based recognition of low resolution faces. In *International Joint Conference onBiometrics*, pages 1 –6, oct. 2011.
132. A. Shrivastava, H. V. Nguyen, V. M. Patel, and R. Chellappa. Design of non-linear discriminative dictionaries for image classification. In *Asian Conference on Computer Vision*, page submitted, 2012.

133. A. Shrivastava, J. K. Pillai, V. M. Patel, and R. Chellappa. Learning discriminative dictionaries with partially labeled data. In *IEEE International Conference on Image Processing*, pages 1–14, 2012.

134. Tal Simchony, Rama Chellappa, and M. Shao. Direct analytical methods for solving poisson equations in computer vision problems. *IEEE Trans. Pattern Anal. Mach. Intell.*, 12(5):435–446, 1990.

135. P. Sinha, B. Balas, Y. Ostrovsky, and R. Russell. Face recognition by humans: Nineteen results all computer vision researchers should know about. *Proceedings of the IEEE*, 94(11):1948 –1962, nov. 2006.

136. M. Soumekh. *Synthetic Aperture Radar Signal Processing With Matlab Algorithms*. Wiley, New York, NY, 1999.

137. R. Tibshirani. Regression shrinkage and selection via the lasso. *Journal of the Royal Statistical Society: Series B*, 58(1):267–288, 1996.

138. J. A. Tropp and A. C. Gilbert. Signal recovery from partial information via orthogonal matching pursuit. *IEEE Transactions on Information Theory*, 53(12):4655–4666, 2006.

139. Joel A. Tropp. Algorithms for simultaneous sparse approximation: part ii: Convex relaxation. *Signal Process.*, 86(3):589–602, March 2006.

140. Joel A. Tropp, Anna C. Gilbert, and Martin J. Strauss. Algorithms for simultaneous sparse approximation: part i: Greedy pursuit. *Signal Process.*, 86(3):572–588, March 2006.

141. J. Trzasko and A. Manduca. Highly undersampled magnetic resonance image reconstruction via homotopic ell_0 -minimization. *IEEE Transactions on Medical Imaging*, 28(1):106 –121, jan. 2009.

142. Y. Tsaig and D. L. Donoho. Extensions of compressed sensing. *Signal Processing*, 86(3):533–548, March 2006.

143. P. Tseng. Convergence of a block coordinate descent method for nondifferentiable minimization. *J. Optimization Theory and Applications*, 109(3):475–494, June 2006.

144. P. Turaga, A. Veeraraghavan, and R. Chellappa. Unsupervised view and rate invariant clustering of video sequences. *CVIU*, 113(3):353–371, 2009.

145. O. Tuzel, F. M. Porikli, and P. Meer. Region covariance: A fast descriptor for detection and classification. In *ECCV*, pages II: 589–600, 2006.

146. N. Vaswani. Kalman filtered compressed sensing. *IEEE International Conference on Image Processing*, (1):893–896, 2008.

147. A. Veeraraghavan, D. Reddy, and R. Raskar. Coded strobing photography: Compressive sensing of high speed periodic videos. *IEEE Transactions on Pattern Analysis and Machine Intelligence*, 33(4):671 –686, 2011.

148. A. Veeraraghavan, A. K. Roy-Chowdhury, and R. Chellappa. Matching shape sequences in video with applications in human movement analysis. *TPAMI*, 27:1896–1909, 2005.

149. M.B. Wakin, J.N. Laska, M.F. Duarte, D. Baron, S. Sarvotham, D. Takhar, K.F. Kelly, and R.G. Baraniuk. An architecture for compressive imaging. In *IEEE International Conference on Image Processing*, pages 1273 –1276, oct. 2006.

150. E. Wang, J. Silva, and L. Carin. Compressive particle filtering for target tracking. *IEEE Workshop on Statistical Signal Processing*, pages 233–236, 2009.

151. Y. Wang, W. Yin, and Y. Zhang. A new alternating minimization algorithm for total variation image reconstruction. *SIAM Journal on Imaging Sciences*, 1(3):248–272, July 2008.

152. G. Warnell, D. Reddy, and R. Chellappa. Adaptive rate compressive sensing for background subtraction. *IEEE International Conference on Acoustics, Speech, and Signal Processing*, 2012.

153. R. M. Willett, R. F. Marcia, and J. M. Nichols. Compressed sensing for practical optical imaging systems: a tutorial. *Optical Engineering*, 50(7):072601(1)–072601(13), Jul. 2011.

154. G.A. Wright. Magnetic resonance imaging. *IEEE Signal Processing Magazine*, 14(1):56–66, jan 1997.

155. J. Wright, Yi Ma, J. Mairal, G. Sapiro, T.S. Huang, and Shuicheng Yan. Sparse representation for computer vision and pattern recognition. *Proceedings of the IEEE*, 98(6):1031 –1044, june 2010.

156. J. Wright, A. Y. Yang, A. Ganesh, S. S. Sastry, and Y. Ma. Robust face recognition via sparse representation. *IEEE Trans. Pattern Analysis and Machine Intelligence*, 31(2):210–227, 2009.
157. S. J. Wright, R. D. Nowak, and M. A. T. Figueiredo. Sparse Reconstruction by Separable Approximation. *preprint*.
158. J. Yang, Y. Zhang, and W. Yin. A fast alternating direction method for tvl1-l2 signal reconstruction from partial fourier data. *IEEE Journal of Selected Topics in Signal Processing*, 4(2):288 –297, april 2010.
159. W. Yin, S. Osher, J. Darbon, and D. Goldfarb. Bregman iterative algorithms for compressed sensing and related problems. Technical report, UCLA CAM.
160. M. Yuan and Y. Lin. Model selection and estimation in regression with grouped variables. *Journal of the Royal Statistical Society: Series B*, 68(1):49–67, 2006.
161. X.-T. Yuan and S. Yan. Visual classification with multi-task joint sparse representation. In *Computer Vision and Pattern Recognition*, 2010.
162. L. Yujiri, M. Shoucri, and P. Moffa. Passive millimeter wave imaging. *IEEE Microwave Magazine*, 4(3):39 – 50, sept. 2003.
163. Li Zhang, Wei-Da Zhou, Pei-Chann Chang, Jing Liu, Zhe Yan, Ting Wang, and Fan-Zhang Li. Kernel sparse representation-based classifier. *IEEE Transactions on Signal Processing*, 60(4):1684 –1695, april 2012.
164. Q. Zhang and B. Li. Discriminative k-svd for dictionary learning in face recognition. In *Computer Vision and Pattern Recognition*, 2010.